Mobile Artificial Intelligence Projects

Develop seven projects on your smartphone using artificial intelligence and deep learning techniques

Karthikeyan NG
Arun Padmanabhan
Matt R. Cole

BIRMINGHAM - MUMBAI

Mobile Artificial Intelligence Projects

Copyright © 2019 Packt Publishing

All rights reserved. No part of this book may be reproduced, stored in a retrieval system, or transmitted in any form or by any means, without the prior written permission of the publisher, except in the case of brief quotations embedded in critical articles or reviews.

Every effort has been made in the preparation of this book to ensure the accuracy of the information presented. However, the information contained in this book is sold without warranty, either express or implied. Neither the authors, nor Packt Publishing or its dealers and distributors, will be held liable for any damages caused or alleged to have been caused directly or indirectly by this book.

Packt Publishing has endeavored to provide trademark information about all of the companies and products mentioned in this book by the appropriate use of capitals. However, Packt Publishing cannot guarantee the accuracy of this information.

Commissioning Editor: Pravin Dhandre
Acquisition Editor: Yogesh Deokar
Content Development Editor: Snehal Kolte
Technical Editor: Naveen Sharma
Copy Editor: Safis Editing
Language Support Editors: Hayden Edwards and Storm Mann
Project Coordinator: Manthan Patel
Proofreader: Safis Editing
Indexer: Tejal Daruwale Soni
Graphics: Jisha Chirayil
Production Coordinator: Shraddha Falebhai

First published: March 2019

Production reference: 1300319

Published by Packt Publishing Ltd.
Livery Place
35 Livery Street
Birmingham
B3 2PB, UK.

ISBN 978-1-78934-407-3

www.packtpub.com

To my wife, Nanthana, for putting up with me during the course of this book. I know it must not have been easy.

To my parents, for their constant support.

`mapt.io`

Mapt is an online digital library that gives you full access to over 5,000 books and videos, as well as industry leading tools to help you plan your personal development and advance your career. For more information, please visit our website.

Why subscribe?

- Spend less time learning and more time coding with practical eBooks and videos from over 4,000 industry professionals

- Improve your learning with Skill Plans built especially for you

- Get a free eBook or video every month

- Mapt is fully searchable

- Copy and paste, print, and bookmark content

Packt.com

Did you know that Packt offers eBook versions of every book published, with PDF and ePub files available? You can upgrade to the eBook version at `www.packt.com` and, as a print book customer, you are entitled to a discount on the eBook copy. Get in touch with us at `customercare@packtpub.com` for more details.

At `www.packt.com`, you can also read a collection of free technical articles, sign up for a range of free newsletters, and receive exclusive discounts and offers on Packt books and eBooks.

Contributors

About the authors

Karthikeyan NG is the Head of Engineering and Technology at the Indian lifestyle & fashion retail brand. He served as a software engineer at Symantec Corporation and has worked with 2 US-based startups as an early employee and has built various products. He has 9+ years of experience in various scalable products using Web, Mobile, ML, AR, and VR technologies. He is an aspiring entrepreneur and technology evangelist. His interests lie in exploring new technologies and innovative ideas to resolve a problem. He has also bagged prizes from more than 15 hackathons, is a TEDx speaker and a speaker at technology conferences and meetups as well as guest lecturer at a Bengaluru University. When not at work, he is found trekking.

> *I would like to extend my gratitude to Manthan Patel for presenting the idea of the book, and to Snehal Kolte for her tenacity. Thanks to Naveen, the technical editor, and the entire editorial team. I would also like to thank the open source community for making this book possible with the frameworks on both Android and iOS platforms.*

Arun Padmanabhan is a machine learning consultant with over 8 years of experience building end-to-end machine learning solutions and applications. Presently working with a couple of start-ups in the financial and insurance sectors, he specializes in automating manual workflows using AI and creating machine vision and NLP applications. Previously, he led the data science team of a Singapore-based product start-up in the restaurant domain. Over the years, he has also built standalone and integrated machine learning solutions in the manufacturing, shipping, and e-commerce domains. His interests lie in research, development, and applications of AI and deep architectures.

Matt R. Cole is a seasoned developer and author with 30 years' experience in Microsoft Windows, C, C++, C# and .Net. Matt is the owner of Evolved AI Solutions, a premier provider of advanced Machine Learning and Bio-AI technologies. Matt is a leading authority on Microservices, and developed the first enterprise grade Microservice framework written completely in C# and .Net. Matt also developed the first Bio Artificial Intelligence framework which completely integrates mirror and canonical neurons. He continues to push the limits of Machine Learning, Biological or Swarm Artificial Intelligence, and Genetic computing.

About the reviewers

Subhash Shah works as head of technology at AIMDek Technologies Pvt. Ltd. He is an experienced solutions architect with over 12 years of experience. He holds a degree in information technology. He is an advocate of open source development and its utilization in solving critical business problems at a reduced cost. His interests include microservices, data analysis, machine learning, AI, and databases. He is an admirer of quality code and **test-driven development** (**TDD**). His technical skills include, but are by no means limited to, translating business requirements into scalable architecture, designing sustainable solutions, and project delivery. He is a coauthor of *MySQL 8 Administrator's Guide* and *Hands-On High Performance with Spring 5*.

Rajib Bhattacharya currently works as a pre-sales consultant with a premier IT solution provider in the Middle East. Rajib has more than 14 years of experience in IT consulting, defining solution architecture, and pre-sales. He is a seasoned architect focusing on analytics, data warehousing, big data, and cognitive computing. He holds multiple patents and is also an author of two books in the field of analytics. He is passionate about helping clients to consolidate their enterprise data, thereby enabling them to make valuable business decisions. Rajib holds a MCA degree from West Bengal University of Technology. He has developed and delivered various training programs on analytics and cognitive computing.

Packt is searching for authors like you

If you're interested in becoming an author for Packt, please visit authors.packtpub.com and apply today. We have worked with thousands of developers and tech professionals, just like you, to help them share their insight with the global tech community. You can make a general application, apply for a specific hot topic that we are recruiting an author for, or submit your own idea.

Table of Contents

Preface

We're witnessing a revolution in **Artificial Intelligence (AI)**, thanks to breakthroughs in deep learning. *Mobile Artificial Intelligence Projects* empowers you to take part in this revolution by applying AI techniques to design applications for **natural language processing (NLP)**, robotics, and computer vision.

The book teaches you how to harness the power of AI in mobile applications, along with learning the core functions of NLP, neural networks, deep learning, and the Mobile Vision API. This book features a range of projects covering tasks such as automated reasoning, facial recognition, digital assistants, automatic text generation, and personalized news and stories. You will learn how to utilize NLP and machine learning algorithms to make applications more predictive, proactive, and capable of making autonomous decisions with less human input. In the concluding chapters, you will work with popular libraries such as TensorFlow Lite, CoreML, and the Snapdragon **Neural Processing Engine (NPE)** across the Android and iOS platforms.

By the end of this book, you will be capable of developing exciting and intuitive mobile applications, delivering a customized and personalized experience to users.

Who this book is for

Mobile Artificial Intelligence Projects is for machine learning professionals, deep learning engineers, AI engineers, and software engineers who want to integrate AI technology into mobile-based platforms and applications.

What this book covers

Chapter 1, *Artificial Intelligence Concepts and Fundamentals*, covers the main concepts and high-level theory required to build AI applications on mobile or the web. We discuss the fundamentals of **Artificial Neural Networks (ANNs)** and deep learning, which form the crux of the current research and trends in AI. We will come to understand all the essential terms required to start our journey on building AI applications.

Chapter 2, *Creating a Real-Estate Price Prediction Mobile App*, is the practical primer for the entire book. We will look at all the essential tools and libraries used throughout the book, and consider how to set up a deep learning environment. We will introduce these by showing how to build a real-estate price prediction app using TensorFlow, deploying it for consumption on mobile and the web. For this, we will be using ANNs and TensorFlow.

Chapter 3, *Implementing Deep Net Architectures to Recognize HandWritten Digits*, focuses on the essential theory, terms, and concepts associated with machine vision. We will get an intuitive understanding of how Convolutional Neural Networks (**CNNs**) work by applying machine vision in practice. We will gain an understanding of how to build real-world applications in machine vision. This chapter will be intuitive and application-focused, instead of theory-heavy.

Chapter 4, *Building a Machine Vision Mobile App to Classify Flower Species*, allows us to translate our learning from the previous chapters to build an object recognition app and then customize it to build an app to classify over 100 species of flowers and pull up their wiki pages. We will learn how to retrain existing Deepnet architectures for custom use cases in object and image classification.

Chapter 5, *Building an ML Model to Predict Car Damage using TensorFlow*, focuses on unsupervised tasks on image restoration using AI. It discusses the deepnets and libraries used for these tasks. We will explore techniques used in deep learning to solve and execute these tasks individually, and then we'll set up and run image restoration from Android and iOS apps.

Chapter 6, *PyTorch Experiments on NLP and RNN*, focuses on the workings of **Recurrent Neural Networks** (**RNNs**) and the applications of AI and NLP. We will deep dive into practically solving NLP use cases in AI.

Chapter 7, *TensorFlow on Mobile with Speech-to-Text with the WaveNet Model*, in this chapter, we are going to learn how to convert audio to text using the WaveNet model. We will then build a model that will take audio and convert it into text using an Android application.

Chapter 8, *Implementing GANs to Recognize Handwritten Digits*, in this chapter, we will build an Android application that detects handwritten numbers and works out what the number is by using adversarial learning. We will use the **Modified National Institute of Standards and Technology** (**MNIST**) dataset for digit classification. We will also look into the basics of **Generative Adversarial Networks** (**GANs**)

Chapter 9, *Sentiment Analysis over Text Using LinearSVC*, in this chapter, we are going to build an iOS application to do sentiment analysis over text and image through user input. We will use existing data models that were built for the same purpose by using LinearSVC, and convert those models into core **machine learning** (**ML**) models for ease of use in our application.

Chapter 10, *What is Next?*, discusses the popular ML-based cloud services and where to start when you build your first ML-based mobile app, and also some references for further reading.

To get the most out of this book

Sound knowledge of machine learning and experience with any programming language is all you need to get started with this book.

Download the example code files

You can download the example code files for this book from your account at www.packt.com. If you purchased this book elsewhere, you can visit www.packt.com/support and register to have the files emailed directly to you.

You can download the code files by following these steps:

1. Log in or register at www.packt.com.
2. Select the **SUPPORT** tab.
3. Click on **Code Downloads & Errata**.
4. Enter the name of the book in the **Search** box and follow the onscreen instructions.

Once the file is downloaded, please make sure that you unzip or extract the folder using the latest version of:

- WinRAR/7-Zip for Windows
- Zipeg/iZip/UnRarX for Mac
- 7-Zip/PeaZip for Linux

The code bundle for the book is also hosted on GitHub at `https://github.com/PacktPublishing/Mobile-Artificial-Intelligence-Projects`. In case there's an update to the code, it will be updated on the existing GitHub repository.

We also have other code bundles from our rich catalog of books and videos available at `https://github.com/PacktPublishing/`. Check them out!

Download the color images

We also provide a PDF file that has color images of the screenshots/diagrams used in this book. You can download it here: `http://www.packtpub.com/sites/default/files/downloads/9781789344073_ColorImages.pdf`.

Conventions used

There are a number of text conventions used throughout this book.

`CodeInText`: Indicates code words in text, database table names, folder names, filenames, file extensions, pathnames, dummy URLs, user input, and Twitter handles. Here is an example: "First, let's import the `math` library so that we can use the `exponential` function."

A block of code is set as follows:

```
def sigmoid ( x ):
    return 1 / ( 1 + e **- x )
```

Any command-line input or output is written as follows:

```
pip install tensorflow
```

Bold: Indicates a new term, an important word, or words that you see onscreen. For example, words in menus or dialog boxes appear in the text like this. Here is an example: "Use the **New** dropdown in the top-right corner to create a new **Python 3** notebook."

 Warnings or important notes appear like this.

 Tips and tricks appear like this.

Get in touch

Feedback from our readers is always welcome.

General feedback: If you have questions about any aspect of this book, mention the book title in the subject of your message and email us at customercare@packtpub.com.

Errata: Although we have taken every care to ensure the accuracy of our content, mistakes do happen. If you have found a mistake in this book, we would be grateful if you would report this to us. Please visit www.packt.com/submit-errata, selecting your book, clicking on the Errata Submission Form link, and entering the details.

Piracy: If you come across any illegal copies of our works in any form on the Internet, we would be grateful if you would provide us with the location address or website name. Please contact us at copyright@packt.com with a link to the material.

If you are interested in becoming an author: If there is a topic that you have expertise in and you are interested in either writing or contributing to a book, please visit authors.packtpub.com.

Reviews

Please leave a review. Once you have read and used this book, why not leave a review on the site that you purchased it from? Potential readers can then see and use your unbiased opinion to make purchase decisions, we at Packt can understand what you think about our products, and our authors can see your feedback on their book. Thank you!

For more information about Packt, please visit packt.com.

Artificial Intelligence Concepts and Fundamentals

1

This chapter acts as a prelude to the entire book and the concepts within it. We will understand these concepts at a level high enough for us to appreciate what we will be building throughout the book.

We will start by getting our head around the general structure of **Artificial Intelligence** (**AI**) and its building blocks by comparing AI, machine learning, and deep learning, as these terms can be used interchangeably. Then, we will skim through the history, evolution, and principles behind **Artificial Neural Networks** (**ANNs**). Later, we will dive into the fundamental concepts and terms of ANNs and deep learning that will be used throughout the book. After that, we take a brief look at the TensorFlow Playground to reinforce our understanding of ANNs. Finally, we will finish off the chapter with thoughts on where to get a deeper theoretical reference for the high-level concepts of the AI and ANN principles covered in this chapter, which will be as follows:

- AI versus machine learning versus deep learning
- Evolution of AI
- The mechanics behind ANNs
- Biological neurons
- Working of artificial neurons
- Activation and cost functions
- Gradient descent, backpropagation, and softmax
- TensorFlow Playground

AI versus machine learning versus deep learning

AI is no new term given the plethora of articles we read online and the many movies based on it. So, before we proceed any further, let's take a step back and understand AI and the terms that regularly accompany it from a practitioner's point of view. We will get a clear distinction of what machine learning, deep learning, and AI are, as these terms are often used interchangeably:

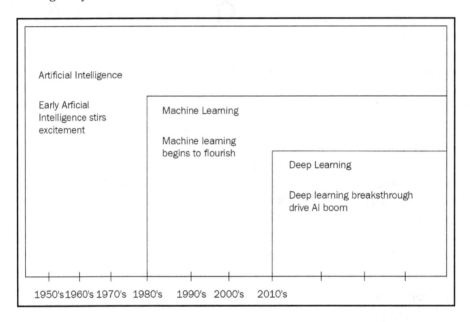

AI is the capability that can be embedded into machines that allows machines to perform tasks that are characteristic of human intelligence. These tasks include seeing and recognizing objects, listening and distinguishing sounds, understanding and comprehending language, and other similar tasks.

Machine learning (**ML**) is a subset of AI that encompasses techniques used to make these human-like tasks possible. So, in a way, ML is what is used to achieve AI.

In essence, if we did not use ML to achieve these tasks, then we would actually be trying to write millions of lines of code with complex loops, rules, and decision trees.

ML gives machines the ability to learn without being explicitly programmed. So, instead of hardcoding rules for every possible scenario to a task, we simply provide examples of how the task is done versus how it should not be done. ML then trains the system on this provided data so it can learn for itself.

ML is an approach to AI where we can achieve tasks such as grouping or clustering, classifying, recommending, predicting, and forecasting data. Some common examples of this are classifying spam mail, stock market predictions, weather forecasting, and more.

Deep learning is a special technique in ML that emulates the human brain's biological structure and works to accomplish human-like tasks. This is done by building a network of neurons just like in the brain through an algorithmic approach using ANNs, which are stack of algorithms that can solve problems at human-like efficiency or better.

These layers are commonly referenced as **deepnets** (deep architectures) and each has a specific problem that it can be trained to solve. The deep learning space is currently at the cutting edge of what we see today, with applications such as autonomous driving, Alexa and Siri, machine vision, and more.

Throughout this book, we will be executing tasks and building apps that are built using these deepnets, and we will also solve use cases by building our very own deepnet architecture.

Evolution of AI

To appreciate what we can currently do with AI, we need to get a basic understanding of how the idea of emulating the human brain was born, and how this idea evolved to a point where we can easily solve tasks in vision and language with human-like capability through machines.

It all started in 1959 when a couple of Harvard scientists, Hubel and Wiesel, were experimenting with a cat's visual system by monitoring the primary visual cortex in the cat's brain.

The **primary visual cortex** is a collection of neurons in the brain placed at the back of the skull and is responsible for processing vision. It is the first part of the brain that receives input signals from the eye, very much like how a human brain would process vision.

The scientists started by showing complex pictures such as those of fish, dogs, and humans to the cat and observed its primary visual cortex. To their disappointment, they got no reading from the primary visual cortex initially. Consequently, to their surprise on one of the trials, as they were removing the slides, dark edges formed, causing some neurons to fire in the primary visual cortex:

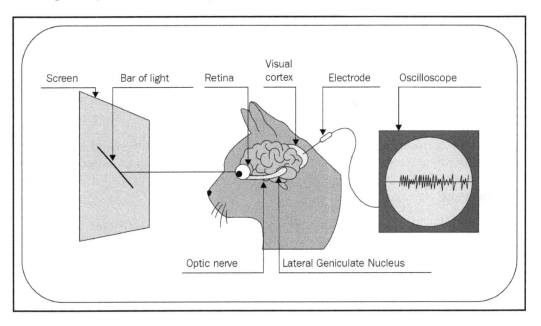

Their serendipitous discovery was that these individual neurons or brain cells in the primary visual cortex were responding to bars or dark edges at various specific orientations. This led to the understanding that the mammalian brain processes a very small amount of information at every neuron, and as the information is passed from neuron to neuron, more complex shapes, edges, curves, and shades are comprehended. So, all these independent neurons holding very basic information need to fire together to comprehend a complete complex image.

After that, there was a lull in the progress of how to emulate the mammalian brain until 1980, when Fukushima proposed neocognitron. **Neocognitron** is inspired by the idea that we should be able to create an increasingly complex representation using a lot of very simplistic representations – just like the mammalian brain!

The following is a representation of how neocognitron works, by Fukushima:

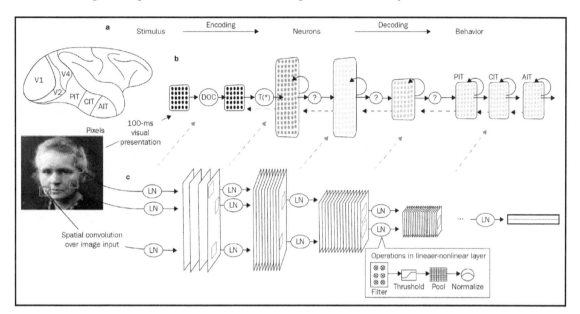

He proposed that to identify your grandmother, there are a lot of neurons that are triggered in the primary visual cortex, and each cell or neuron understands an abstract part of the final image of your grandmother. All of these neurons work in sequence, parallel, and tandem, and then finally hits a grandmother cell or neuron which fires only when it sees your grandmother.

Fast forward to today (2010-2018), with contributions from Yoshua Bengio, Yann LeCun, and Geoffrey Hinton, who are commonly known as the *fathers of deep learning*. They contribute massively to the AI space we work in today. They have given rise to a whole new approach to machine learning where feature engineering is automated.

The idea of not explicitly telling the algorithm what it should be looking for and letting it figure this out by itself by feeding it a lot of examples is the latest development. The analogy to this principle would be that of teaching a child to distinguish between an apple and an orange. We would show the child pictures of apples and oranges rather than only describing the two fruits' features, such as shape, color, size, and so on.

The following diagram shows the difference between ML and deep learning:

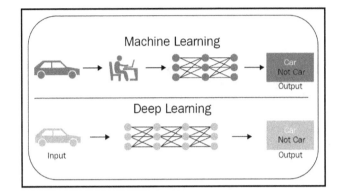

This is the primary difference between traditional ML and ML using neural networks (deep learning). In traditional ML, we provide features along with labels, but using ANNs, we let the algorithm decipher the features.

We live in an exciting time, an era we share with the fathers of deep learning, so much so that there are exchanges online in places such as Stack Exchange, where we can see contributions even from Yann LeCun and Geoffrey Hinton. This is analogous to living in the time of, and writing to, Nicholas Otto, the father of the internal combustion engine, who started the automobile revolution that we see evolving even to this day. The automobile revolution will be dwarfed by what could be possible with AI in the future. Exciting times, indeed!

The mechanics behind ANNs

In this section, we will understand the nuts and bolts that are required to start building our own AI projects. We will get to grips with the common terms that are used in deep learning techniques.

This section aims to provide the essential theory at a high level, giving you enough insight so that you're able to build your own deep neural networks, tune them, and understand what it takes to make state-of-the-art neural networks.

Biological neurons

We previously discussed how the biological brain has been an inspiration behind ANNs. The brain is made up of hundreds of billions of independent units or cells called **neurons**.

The following diagram depicts a **neuron**, and it has multiple inputs going into it, called **dendrites**. There is also an output going out of the cell body, called the **axon**:

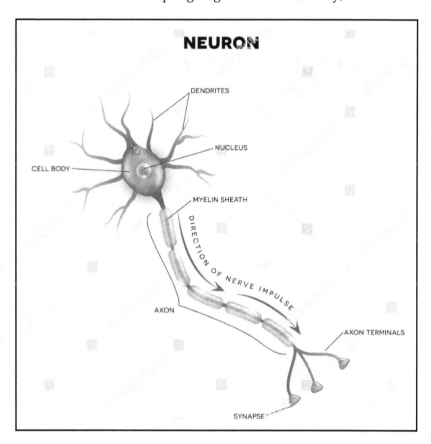

The dendrites carry information into the neuron and the axon allows the processed information to flow out of the neuron. But in reality, there are thousands of dendrites feeding input into the neuron body as small electrical charges. If these small electrical charges that are carried by the dendrites have an effect on the overall charge of the body or cross over some threshold, then the axon will fire.

Now that we know how a biological neuron functions, we will understand how an artificial neuron works.

Working of artificial neurons

Just like the biological brain, ANNs are made up of independent units called neurons. Like the biological neuron, the artificial neuron has a body that does some computation and has many inputs that are feeding into the cell body or neuron:

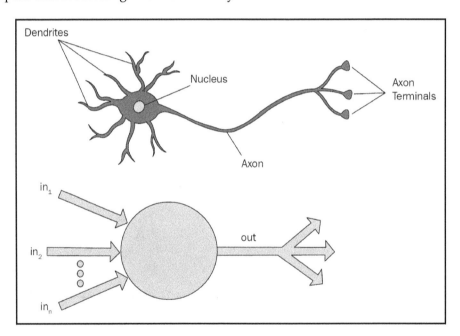

For example, let's assume we have three inputs to the neuron. Each input carries a binary value of 0 or 1. We have an output flowing out of the body, which also carries a binary value of 0 or 1. For this example, the neuron decides whether I should eat a cake today or not. That is, the neuron should fire an output of 1 if I should eat a cake or fire 0 if I shouldn't:

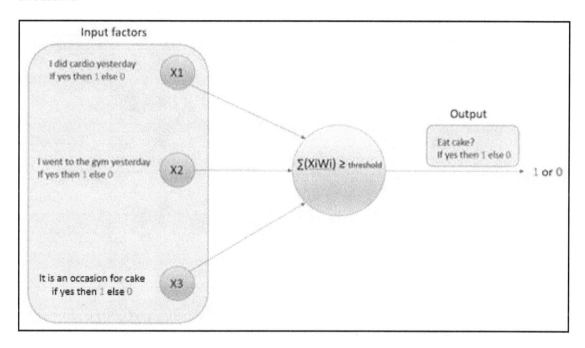

In our example, the three inputs represent the three factors that determine whether I should eat the cake or not. Each factor is given a weight of importance; for instance, the first factor is **I did cardio yesterday** and it has a weight of 2. The second factor is **I went to the gym yesterday** and weighs 3. The third factor is **It is an occasion for cake** and weighs 6.

The body of the neuron does some calculation to inputs, such as taking the sum of all of these inputs and checking whether it is over some threshold:

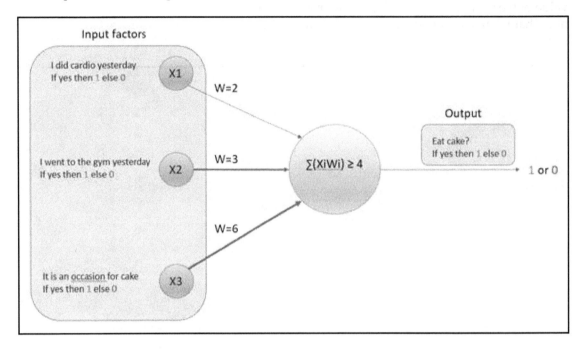

So, for this example, let's set our threshold as 4. If the sum of the input weights is above the threshold, then the neuron fires an output of 1, indicating that I can eat the cake.

This can be expressed as an equation:

$$output \begin{cases} 1 & IF \sum(WiXi + WiiXii + WiiiXiii) \geq threshold \\ 0 & IF \sum(WiXi + WiiXii + WiiiXiii) < threshold \end{cases}$$

Here, the following applies:

- Xi is the first input factor, *I did cardio yesterday.*
- Wi is the weight of the first input factor, Xi. In our example, $Wi = 2$.
- Xii is the second input factor, *I went to the gym yesterday.*
- Wii is the weight of the second input factor, Xii. In our example, $Wii = 3$.
- $Xiii$ is the third input factor, *It is an occasion for cake.*
- $Wiii$ is the weight of the third input factor, $Xiii$. In our example, $Wiii = 6$.
- *threshold* is 4.

Now, let's use this neuron to decide whether I can eat a cake for three different scenarios.

Scenario 1

I want to eat a cake and I went to the gym yesterday, but I did not do cardio, nor is it an occasion for cake:

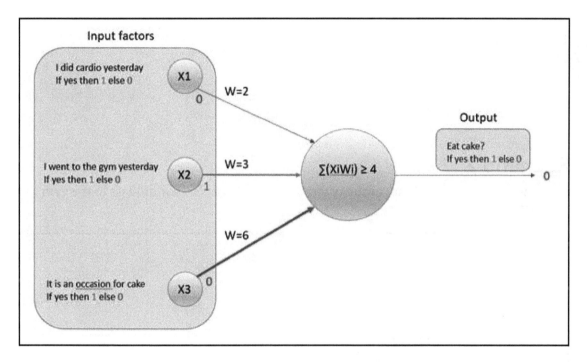

Here, the following applies:

- Xi is the first input factor, *I did cardio yesterday*. Now, $Xi = 0$, as this is false.
- Wi is the weight of the first input factor, Xi. In our example, $Wi = 2$.
- Xii is the second input factor, *I went to the gym yesterday*. Now, $Xii = 1$, as this is true.
- Wii is the weight of the second input factor, Xii. In our example, $Wii = 3$.
- $Xiii$ is the third input factor, *It is an occasion for cake*. Now, $Xiii = 0$, as this is false.
- $Wiii$ is the weight of the third input factor, $Xiii$. In our example, $Wiii = 6$.
- *threshold* is 4.

We know that the neuron computes the following equation:

$$\sum (WiXi + WiiXii + WiiiXiii) \geq threshold$$

For scenario 1, the equation will translate to this:

$$\sum (2 * 0 + 3 * 1 + 6 * 0) \geq 4$$

This is equal to this:

$$\sum (0 + 3 + 0) \geq 4$$

$3 \geq 4$ is false, so it fires 0, which means I should not eat the cake.

Scenario 2

I want to eat a cake and it's my birthday, but I did not do cardio, nor did I go to the gym yesterday:

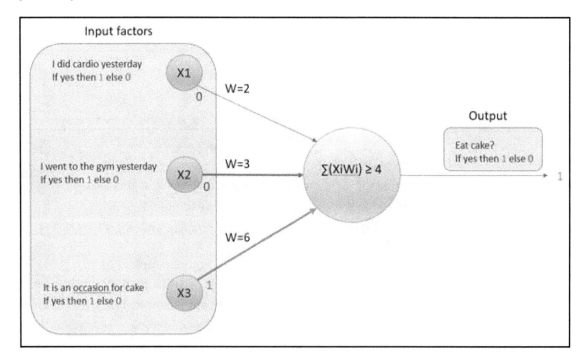

Here, the following applies:

- *Xi* is the first input factor, *I did cardio yesterday*. Now, *Xi* = 0, as this factor is false.
- *Wi* is the weight of the first input factor, *Xi*. In our example, *Wi* = 2.
- *Xii* is the second input factor, *I went to the gym yesterday*. Now, *Xii* = 0, as this factor is false.
- *Wii* is the weight of the second input factor, *Xii*. In our example, *Wii* = 3.
- *Xiii* is the third input factor, *It is an occasion for cake*. Now, *Xiii* = 1, this factor is true.
- *Wiii* is the weight of the third input factor, *Xiii*. In our example, *Wiii* = 6.
- *threshold* is 4.

We know that the neuron computes the following equation:

$$\sum (WiXi + WiiXii + WiiiXiii) \geq threshold$$

For scenario 2, the equation will translate to this:

$$\sum (2 * 0 + 3 * 0 + 6 * 1) \geq 4$$

It gives us the following output:

$$\sum (0 + 0 + 6) \geq 4$$

6 ≥ 4 is true, so this fires 1, which means I can eat the cake.

Scenario 3

I want to eat a cake and I did cardio and went to the gym yesterday, but it is also not an occasion for cake:

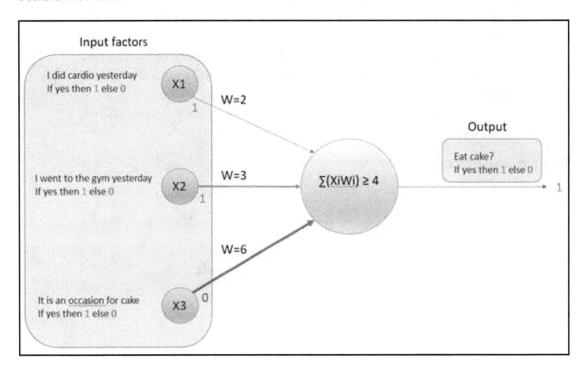

Here, the following applies:

- Xi is the first input factor, *I did cardio yesterday*. Now, $Xi = 1$, as this factor is true.
- Wi is the weight of the first input factor, Xi. In our example, $Wi = 2$.
- Xii is the second input factor, *I went to the gym yesterday*. Now, $Xii = 1$, as this factor is true.
- Wii is the weight of the second input factor, Xii. In our example, $Wii = 3$.
- $Xiii$ is the third input factor, *It is an occasion for cake*. Now, $Xiii = 0$, as this factor is false.
- $Wiii$ is the weight of the third input factor, $Xiii$. In our example, $Wiii = 6$.
- *threshold* is 4.

We know that the neuron computes the following equation:

$$\sum (WiXi + WiiXii + WiiiXiii) \geq threshold$$

For scenario 3, the equation will translate to this:

$$\sum (2*1 + 3*1 + 6*0) \geq 4$$

This gives us the following equation:

$$\sum (2 + 3 + 0) \geq 4$$

$5 \geq 4$ is true, so this fires 1, which means I can eat the cake.

From the preceding three scenarios, we saw how a single artificial neuron works. This single unit is also called a **perceptron**. A perceptron essentially handles binary inputs, computes the sum, and then compares with a threshold to ultimately give a binary output.

To better appreciate how a perceptron works, we can translate our preceding equation into a more generalized form for the sake of explanation.

Let's assume there is just one input factor, for simplicity:

$$\sum (WiXi + WiiXii + WiiiXiii) = w.x$$

Let's also assume that *threshold = b*. Our equation was as follows:

$$output \begin{cases} 1 & IF \sum (WiXi + WiiXii + WiiiXiii) \geq threshold \\ 0 & IF \sum (WiXi + WiiXii + WiiiXiii) < threshold \end{cases}$$

It now becomes this:

$$output \begin{cases} 1 & IF(w.x \geq b) \\ 0 & IF(w.x < b) \end{cases}$$

It can also be written as $IF\ w.x + b \geq 0$, then output *1* else *0*.

Here, the following applies:

- w is the weight of the input
- b is the threshold and is referred to as the bias

This rule summarizes how a perceptron neuron works.

Just like the mammalian brain, an ANN is made up of many such perceptions that are stacked and layered together. In the next section, we will get an understanding of how these neurons work together within an ANN.

ANNs

Like biological neurons, artificial neurons also do not exist on their own. They exist in a network with other neurons. Basically, the neurons exist by feeding information to each other; the outputs of some neurons are inputs to some other neurons.

In any ANN, the first layer is called the **Input Layer**. These inputs are real values, such as the factors with weights (*w.x*) in our previous example. The sum values from the input layer are propagated to each neuron in the next layer. The neurons of that layer do the computation and pass their output to the next layer, and so on:

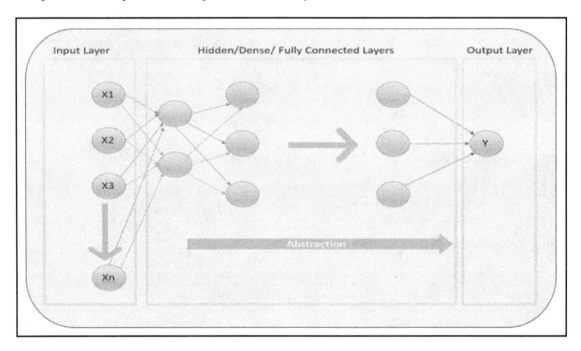

The layer that receives input from all previous neurons and passes its output to all of the neurons of the next layer is called a **Dense** layer. As this layer is connected to all of the neurons of the previous and next layer, it is also commonly referred to as a **Fully Connected Layer**.

The input and computation flow from layer to layer and finally end at the **Output Layer**, which gives the end estimate of the whole ANN.

The layers in-between the input and the output layers are called the **Hidden Layers**, as the values of the neurons within these hidden layers are unknown and a complete black box to the practitioner.

As you increase the number of layers, you increase the abstraction of the network, which in turn increases the ability of the network to solve more complex problems. When there are over three hidden layers, then it is referred to as a deepnet.

So, if this was a machine vision task, then the first hidden layer would be looking for edges, the next would look for corners, the next for curves and simple shapes, and so on:

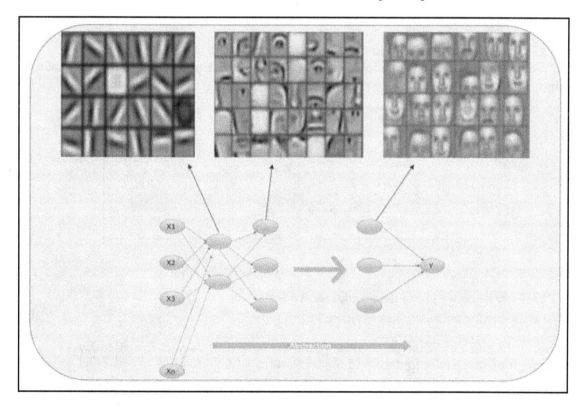

Therefore, the complexity of the problem can determine the number of layers that are required; more layers lead to more abstractions. These layers can be very deep, with 1,000 or more layers, to very shallow, with just about half a dozen layers. Increasing the number of hidden layers does not necessarily give better results as the abstractions may be redundant.

So far, we have seen how artificial neurons can be stacked together to form a neural network. But we have seen that the perceptron neuron takes only binary input and gives only binary output. But in practice, there is a problem in doing things based on the perceptron's idea. This problem is addressed by activation functions.

Activation functions

We now know that an ANN is created by stacking individual computing units called perceptrons. We have also seen how a perceptron works and have summarized it as *Output 1, IF* $w.x + b \geq 0$.

That is, it either outputs a *1* or a *0* depending on the values of the weight, *w*, and bias, *b*.

Let's look at the following diagram to understand why there is a problem with just outputting either a *1* or a *0*. The following is a diagram of a simple perceptron with just a single input, *x*:

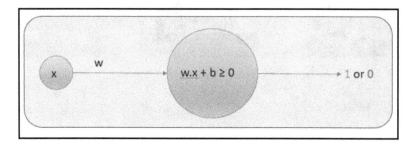

For simplicity, let's call $w.x + b = z$, where the following applies:

- *w* is the weight of the input, *x*, and *b* is the bias
- *a* is the output, which is either *1* or *0*

Here, as the value of *z* changes, at some point, the output, *a*, changes from *0* to *1*. As you can see, the change in output *a* is sudden and drastic:

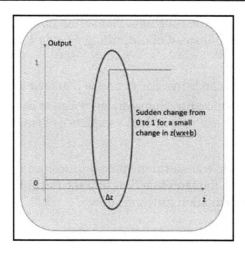

What this means is that for some small change, $\triangle z$, we get a dramatic change in the output, *a*. This is not particularly helpful if the perceptron is part of a network, because if each perceptron has such drastic change, it makes the network unstable and hence the network fails to learn.

Therefore, to make the network more efficient and stable, we need to slow down the way each perceptron learns. In other words, we need to eliminate this sudden change in output from *0* to *1* to a more gradual change:

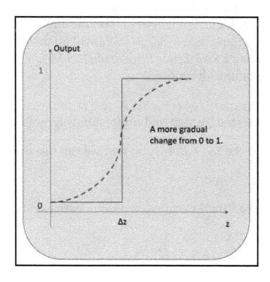

This is made possible by activation functions. Activation functions are functions that are applied to a perceptron so that instead of outputting a 0 or a 1, it outputs any value between 0 and 1.

This means that each neuron can learn slower and at a greater level of detail by using smaller changes, $\triangle z$. Activation functions can be looked at as transformation functions that are used to transform binary values in to a sequence of smaller values between a given minimum and maximum.

There are a number of ways to transform the binary outcomes to a sequence of values, namely the sigmoid function, the tanh function, and the ReLU function. We will have a quick look at each of these activation functions now.

Sigmoid function

The **sigmoid function** is a function in mathematics that outputs a value between 0 and 1 for any input:

$$\Theta(z) = 1/(1 + e^{-z})$$

Here, $z = wx + b$ and $0 > \Theta(z) \leq 1$.

Let's understand sigmoid functions better with the help of some simple code. If you do not have Python installed, no problem: we will use an online alternative for now at `https://www.jdoodle.com/python-programming-online`. We will go through a complete setup from scratch in `Chapter 2`, *Creating a Real-Estate Price Prediction Mobile App*. Right now, let's quickly continue with the online alternative.

Once we have the page at `https://www.jdoodle.com/python-programming-online` loaded, we can go through the code step by step and understand sigmoid functions:

1. First, let's import the `math` library so that we can use the exponential function:

```
from    math    import   e
```

2. Next, let's define a function called `sigmoid`, based on the earlier formula:

```
def sigmoid ( x ):
    return 1 / ( 1 + e **- x )
```

3. Let's take a scenario where our z is very small, -10. Therefore, the function outputs a number that is very small and close to 0:

```
sigmoid(-10)
4.539786870243442e-05
```

4. If z is very large, such as 10000, then the function will output the maximum possible value, 1:

```
sigmoid(10000)
1.0
```

Therefore, the sigmoid function transforms any value, z, to a value between 0 and 1. When the sigmoid activation function is used on a neuron instead of the traditional perceptron algorithm, we get what is called a **sigmoid neuron**:

Tanh function

Similar to the sigmoid neuron, we can apply an activation function called tanh(z), which transforms any value to a value between -1 and 1.

The neuron that uses this activation function is called a **tanh neuron**:

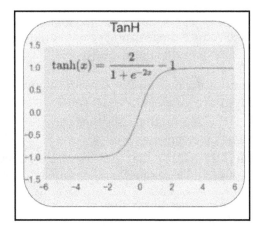

ReLU function

Then there is an activation function called the **Rectified Linear Unit, ReLU(z)**, that transforms any value, z, to 0 or a value above 0. In other words, it outputs any value below 0 as 0 and any value above 0 as the value itself:

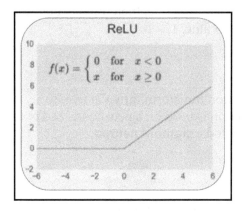

Just to summarize our understanding so far, the perceptron is the traditional and outdated neuron that is rarely used in real implementations. They are great to get a simplistic understanding of the underlying principle; however, they had the problem of fast learning due to the drastic changes in output values.

We use activation functions to reduce the learning speed and determine finer changes in z or $w.x + b$. Let's sum up these activation functions:

- The **sigmoid neuron** is the neuron that uses the sigmoid activation function to transform the output to a value between 0 and 1.
- The **tanh neuron** is the neuron that uses the tanh activation function to transform the output to a value between -1 and 1.
- The **ReLU neuron** is the neuron that uses the ReLU activation function to transform the output to a value of either 0 or any value above 0.

The sigmoid function is used in practice but is slow compared to the tanh and ReLU functions. The tanh and ReLU functions are commonly used activation functions. The ReLU function is also considered state of the art and is usually the first choice of activation function that's used to build ANNs.

Here is a list of commonly used activation functions:

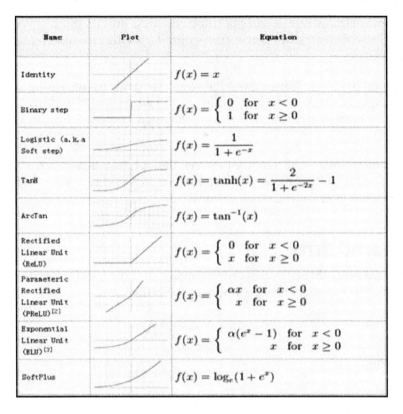

Name	Plot	Equation
Identity		$f(x) = x$
Binary step		$f(x) = \begin{cases} 0 & \text{for } x < 0 \\ 1 & \text{for } x \geq 0 \end{cases}$
Logistic (a.k.a Soft step)		$f(x) = \dfrac{1}{1 + e^{-x}}$
Tanh		$f(x) = \tanh(x) = \dfrac{2}{1 + e^{-2x}} - 1$
ArcTan		$f(x) = \tan^{-1}(x)$
Rectified Linear Unit (ReLU)		$f(x) = \begin{cases} 0 & \text{for } x < 0 \\ x & \text{for } x \geq 0 \end{cases}$
Parameteric Rectified Linear Unit (PReLU) [2]		$f(x) = \begin{cases} \alpha x & \text{for } x < 0 \\ x & \text{for } x \geq 0 \end{cases}$
Exponential Linear Unit (ELU) [3]		$f(x) = \begin{cases} \alpha(e^x - 1) & \text{for } x < 0 \\ x & \text{for } x \geq 0 \end{cases}$
SoftPlus		$f(x) = \log_e(1 + e^x)$

In the projects within this book, we will be primarily using either the sigmoid, tanh, or the ReLU neurons to build our ANN.

Cost functions

To quickly recap, we know how a basic perceptron works and its pitfalls. We then saw how activation functions overcame the perceptron's pitfalls, giving rise to other neuron types that are in use today.

Now, we are going to look at how we can tell when the neurons are wrong. For any type of neuron to learn, it needs to know when it outputs the wrong value and by what margin. The most common way to measure how wrong the neural network is, is to use a cost function.

A **cost function** quantifies the difference between the output we get from a neuron to an output that we need from that neuron. There are two common types of cost functions that are used: mean squared error and cross entropy.

Mean squared error

The **mean squared error** (**MSE**) is also called a quadratic cost function as it uses the squared difference to measure the magnitude of the error:

$$MSE = \sum (y - a)^2 / n$$

Here, the following applies:

- a is the output from the ANN
- y is the expected output
- n is the number of samples used

The cost function is pretty straightforward. For example, consider a single neuron with just one sample, ($n=1$). If the expected output is 2 ($y=2$) and the neuron outputs 3 ($a=3$), then the MSE is as follows:

$$MSE = \sum (2 - 3)^2 / 1$$

$$MSE = (-1)^2$$

$$MSE = 1$$

Similarly, if the expected output is 3 (*y=3*) and the neuron outputs 2 (*a=2*), then the MSE is as follows:

$$MSE = \sum (3-2)^2/1$$

$$MSE = (1)^2$$

$$MSE = 1$$

Therefore, the MSE quantifies the magnitude of the error made by the neuron. One of the issues with MSE is that when the values in the network get large, the learning becomes slow. In other words, when the weights (*w*) and bias (*b*) or z get large, the learning becomes very slow. Keep in mind that we are talking about thousands of neurons in an ANN, which is why the learning slows down and eventually stagnates with no further learning.

Cross entropy

Cross entropy is a derivative-based function as it uses the derivative of a specially designed equation, which is given as follows:

$$CrossEntropy = 1/n * (y * Ln\,(a) + (1-y) * Ln\,(1-a))$$

Cross entropy allows the network to learn faster when the difference between the expected and actual output is greater. In other words, the bigger the error, the faster it helps the network learn. We will get our heads around this using some simple code.

Like before, for now, you can use an online alternative if you do not have Python already installed, at `https://www.jdoodle.com/python-programming-online`. We will cover the installation and setup in `Chapter 2`, *Creating a Real-Estate Price Prediction Mobile App*. Follow these steps to see how a network learns using cross entropy:

1. First, let's import the `math` library so that we can use the `log` function:

   ```
   from numpy import log
   ```

2. Next, let's define a function called `cross_enrtopy`, based on the preceding formula:

   ```
   def cross_entropy(y,a):
       return -1 *(y*log(a)+(1-y)*log (1-a))
   ```

3. For example, consider a single neuron with just one sample, (*n=1*). Say the expected output is 0 (*y=0*) and the neuron outputs 0.01 (*a=0.01*):

```
cross_entropy(0, 0.01)
```

The output is as follows:

```
0.010050335853501451
```

Since the expected and actual output values are very small, the resultant cost is very small.

Similarly, if the expected and actual output values are very large, then the resultant cost is still small:

```
cross_entropy(1000,999.99)
```

The output is as follows:

```
0.010050335853501451
```

Similarly, if the expected and actual output values are far apart, then the resultant cost is large:

```
cross_entropy(0,0.9)
```

The output is as follows:

```
2.3025850929940459
```

Therefore, the larger the difference in expected versus actual output, the faster the learning becomes. Using cross entropy, we can get the error of the network, and at the same time, the magnitude of the weights and bias is irrelevant, helping the network learn faster.

Gradient descent

Up until now, we have covered the different kind of neurons based on the activation functions that are used. We have covered the ways to quantify inaccuracy in the output of a neuron using cost functions. Now, we need a mechanism to take that inaccuracy and remedy it.

The mechanism through which the network can learn to output values closer to the expected or desired output is called **gradient descent**. Gradient descent is a common approach in machine learning for finding the lowest cost possible.

To understand gradient descent, let's use the single neuron equation we have been using so far:

$$w . x + b = z$$

Here, the following applies:

- x is the input
- w is the weight of the input
- b is the bias of the input

Gradient descent can be represented as follows:

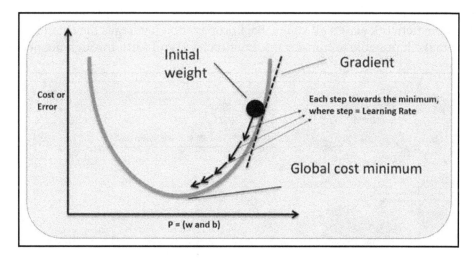

Initially, the neuron starts by assigning random values for w and b. From that point onward, the neuron needs to adjust the values of w and b so that it lowers or decreases the error or cost (cross entropy).

Taking the derivative of the cross entropy (cost function) results in a step-by-step change in w and b in the direction of the lowest cost possible. In other words, **gradient descent** tries to find the finest line between the network output and expected output.

The weights are adjusted based on a parameter called the **learning rate.** The learning rate is the value that is adjusted to the weight of the neuron to get an output closer to the expected output.

Keep in mind that here, we have used only a single parameter; this is only to make things easier to comprehend. In reality, there are thousands upon millions of parameters that are taken into consideration to lower the cost.

Backpropagation – a method for neural networks to learn

Great! We have come a long way, from looking at the biological neuron, to the types of neuron, to determining accuracy, and correcting the learning of the neuron. Only one question remains: *how can the whole network of neurons learn together?*

Backpropagation is an incredibly smart approach to making gradient descent happen throughout the network across all layers. Backpropagation leverages the chain rule from calculus to make it possible to transfer information back and forth through the network:

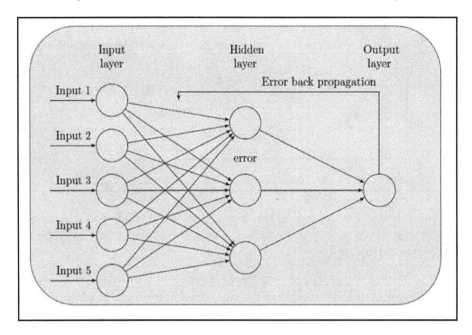

In principle, the information from the input parameters and weights is propagated through the network to make a guess at the expected output and then the overall inaccuracy is backpropagated through the layers of the network so that the weights can be adjusted and the output can be guessed again.

This single cycle of learning is called a **training step** or **iteration**. Each iteration is performed on a batch of the input training samples. The number of samples in a batch is called **batch size**. When all of the input samples have been through an iteration or training step, then it is called an **epoch**.

For example, let's say there are 100 training samples and in every iteration or training step, there are 10 samples being used by the network to learn. Then, we can say that the batch size is 10 and it will take 10 iterations to complete a single epoch. Provided each batch has unique samples, that is, if every sample is used by the network at least once, then it is a single epoch.

This back-and-forth propagation of the predicted output and the cost through the network is how the network learns.

We will revisit training step, epoch, learning rate, cross entropy, batch size, and more during our hands-on sections.

Softmax

We have reached our final conceptual topic for this chapter. We've covered types of neurons, cost functions, gradient descent, and finally a mechanism to apply gradient descent across the network, making it possible to learn over repeated iterations.

Previously, we saw the input layer and dense or hidden layers of an ANN:

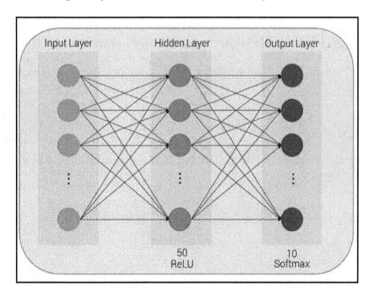

Softmax is a special kind of neuron that's used in the output layer to describe the probability of the respective output:

$$Softmax = e^M i / \sum (e^M j)$$

To understand the softmax equation and its concepts, we will be using some code. Like before, for now, you can use any online Python editor to follow the code.

First, import the exponential methods from the math library:

```
from math import exp
```

For the sake of this example, let's say that this network is designed to classify three possible labels: A, B, and C. Let's say that there are three signals going into the softmax from the previous layers (-1, 1, 5):

```
a=[-1.0,1.0,5.0]
```

The explanation is as follows:

- The first signal indicates that the output should be A, but is weak and is represented with a value of -1
- The second signal indicates that the output should be B and is slightly stronger and represented with a value of 1
- The third signal is the strongest, indicating that the output should be C and is represented with a value of 5

These represented values are confidence measures of what the expected output should be.

Now, let's take the numerator of the softmax for the first signal, guessing that the output is A:

$$Softmax(numerator) = e^M i$$

Here, *M* is the output signal strength indicating that the output should be A:

```
exp(a[0]) # taking the first element of a[-1,1,5] which represents A

0.36787944117144233
```

Next, there's the numerator of the softmax for the second signal, guessing that the output is B:

$$Softmax(numerator) = e^M i$$

Here, M is the output signal strength indicating that the output should be B:

```
exp(a[0]) # taking the second element of a[-1,1,5] which represents B

2.718281828459045
```

Finally, there's the numerator of the softmax for the second signal, guessing that the output is C:

$$Softmax(numerator) = e^M i$$

Here, M is the output signal strength indicating that the output should be C:

```
exp(a[2])
# taking the third element of a[-1,1,5] which represents C

148.4131591025766
```

We can observe that the represented confidence values are always placed above 0 and that the resultant is made exponentially larger.

Now, let's interpret the denominator of the softmax function, which is a sum of the exponential of each signal value:

$$Softmax(denominator) = \sum(e^M j)$$

Let's write some code for softmax function:

```
sigma = exp ( a [ 0 ]) + exp ( a [ 1 ]) + exp ( a [ 2 ])
sigma

151.49932037220708
```

Therefore, the probability that the first signal is correct is as follows:

```
exp(a[0])/sigma

0.0024282580295913376
```

This is less than a 1% chance that it is A.

Similarly, the probability that the third signal is correct is as follows:

```
exp(a[2])/sigma
```

```
0.9796292071670795
```

This means there is over a 97% chance that the expected output is indeed c.

Essentially, the softmax accepts a weighted signal that indicates the confidence of some class prediction and outputs a probability score between 0 to 1 for all of those classes.

Great! We have made it through the essential high-level theory that's required to get us hands on with our projects. Next up, we will summarize our understanding of these concepts by exploring the TensorFlow Playground.

TensorFlow Playground

Before we get started with the TensorFlow Playground, let's recap the essential concepts quickly. It will help us appreciate the TensorFlow Playground better.

The inspiration for neural networks is the biological brain, and the smallest unit in the brain is a **neuron**.

A **Perceptron** is a neuron based on the idea of the biological neuron. The perceptron basically deals with binary inputs and outputs, making it impractical for actual pragmatic purposes. Also, because of its binary nature, it learns too fast due to the drastic change in output for a small change in input, and so does not provide fine details.

Activation functions were used to negate the issue with perceptrons. This gave rise to other types of neurons that deal with values between ranges of 0 to 1, -1 to 1, and so on, instead of just a 0 or a 1.

ANNs are made up of these neurons stacked in layers. There is an input layer, a dense or fully connected layer, and an output layer.

Cost functions, such as MSE and cross entropy, are ways to measure the magnitude of error in the output of a neuron.

Gradient descent is a mechanism through which a neuron can learn to output values closer to the expected or desired output.

Backpropagation is an incredibly smart approach to making gradient descent happen throughout the network across all layers.

Each back and forth propagation or iteration of the predicted output and the cost through the network is called a **training step**.

The **learning rate** is the value that is adjusted to the weight of the neuron at each training step to get an output that's closer to the expected output.

Softmax is a special kind of neuron that accepts a weighted signal indicating the confidence of some class prediction and outputting a probability score between 0 to 1 for all of those classes.

Now, we can proceed to TensorFlow Playground at `https://Playground.tensorflow.org`. TensorFlow Playground is an online tool to visualize an ANN or deepnet in action, and is an excellent place to reiterate what we have learned conceptually in a visual and intuitive way.

Now, without further ado, let's get on with TensorFlow Playground. Once the page is loaded, you will see a dashboard to create your own neural network for predefined classification problems. Here is a screenshot of the default page and its sections:

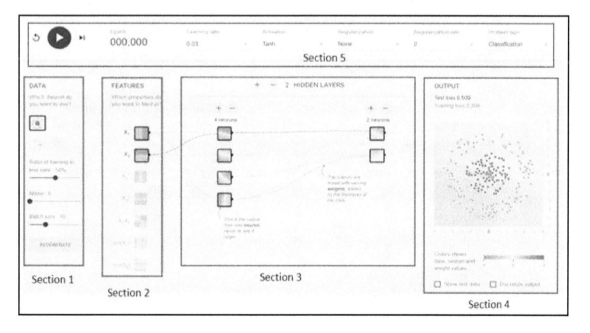

Let's look at each of the sections from this screenshot:

- **Section 1**: The data section shows choices of the pre-built problems to build and visualize the network. The first problem is chosen, which is basically to distinguish between the blue an orange dots. Below that, there are controls to divide the data into training and testing subsets. There is also a parameter to set the batch size. The **Batch size** is the number of samples that are taken into the network for learning during each training step.
- **Section 2**: The features section indicates the number of input parameters. In this case, there are two features chosen as the input features.
- **Section 3**: The hidden layer section is where we can create hidden layers to increase complexity. There are also controls to increase and decrease the number of neurons within each hidden or dense layer. In this example, there are two hidden layers with four and two neurons, respectively.
- **Section 4**: The output section is where we can see the loss or the cost graph, along with a visualization of how well the network has learned to separate the red and blue dots.
- **Section 5**: This section is the control panel for adjusting the tuning parameters of the network. It has a widget to start, pause, and refresh the training of the network. Next to it, there is a counter indicating the number of epochs elapsed. Then there is **Learning rate**, the constant by which the weights are adjusted. That is followed by the choice of activation function to use within the neurons. Finally, there is an option to indicate the kind of problem to visualize, that is classification, or regression. In this example, we are visualizing a classification task.
- We will ignore the **Regularization** and **Regularization rate** for now, as we have not covered these terms in a conceptual manner as of yet. We will visit these terms in later in the book when it is ideal for appreciating its purpose.

We are now ready to start fiddling around with TensorFlow Playground. We will start with the first dataset, with the following settings on the tuning parameters:

- **Learning rate = 0.01**
- **Activation = Tanh**
- **Regularization = None**
- **Regularization rate = 0**
- **Problem type = Classification**
- **DATA = Circle**
- **Ratio of training to test data = 50%**
- **Batch size = 10**

- **FEATURES = X_1 and X_2**
  - Two hidden/dense layers; the first layer with **4 neurons**, and the second layer with **2 neurons**

Now start training by clicking the play button on the top-left corner of the dashboard. Moving right from the play/pause button, we can see the number of epochs that have elapsed. At about 200 epochs, pause the training and observe the output section:

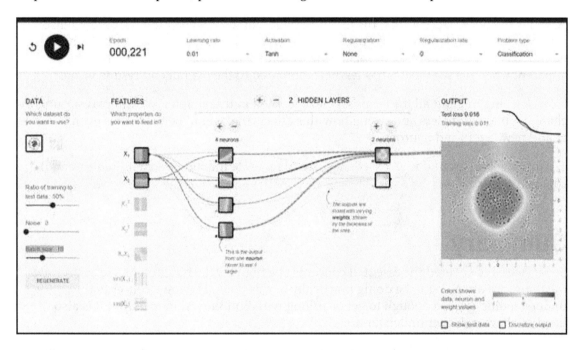

The key observations from the dashboard are as follows:

- We can see the performance graph of the network on the right section of the dashboard. The test and training loss is the cost of the network during testing and training, respectively. As discussed previously, the idea is to minimize cost.
- Below that, you will observe that there is a visualization of how the network has separated or classified the blue dots from the orange ones.
- If we hover the mouse pointer over any of the neurons, we can see what the neuron has learned to separate the blue and orange dots. Having said this, let's take a closer look at both of the neurons from the second layer to see what they have learned about the task.

- When we hover over the first neuron in the second layer, we can see that this neuron has done a good job of learning the task at hand. In comparison, the second neuron in the second layer has learned less about the task.
- That brings us to the dotted lines coming out of the neurons: they are the corresponding weights of the neuron. The blue dotted lines indicate positive weights while the orange dotted ones indicate negative weights. They are commonly called **tensors.**

Another key observation is that the first neuron in the second layer has a stronger tensor signal coming out of it compared to the second one. This is indicative of the influence this neuron has in the overall task of separating the blue and orange dots, and it is quite apparent when we see what it has learned compared to the overall end results visual.

Now, keeping in mind all the terms we have learned in this chapter, we can play around by changing the parameters and seeing how this affects the overall network. It is even possible to add new layers and neurons.

TensorFlow Playground is an excellent place to reiterate the fundamentals and essential concepts of ANNs.

Summary

So far, we have covered the essential concepts at a high level, enough for us to appreciate the things we are going to be doing practically in this book. Having a conceptual understanding is good enough to get us rolling with building AI models, but it is also handy to have a deeper understanding.

In the next chapter, we will set up our environment for building AI applications and create a small Android and iOS mobile app that can use a model built on Keras and TensorFlow to predict house prices.

Further reading

Here is a list of resources that can be referenced to appreciate and dive deeper into the concepts of AI and deep learning:

- *Neural Networks and deep learning,* http://neuralnetworksanddeeplearning.com/
- Michael Taylor's *Make Your Own Neural Network: An In-depth Visual Introduction For Beginners,* https://www.amazon.in/Machine-Learning-Neural-Networks-depth-ebook/dp/B075882XCP
- Tariq Rashid's *Make Your Own Neural Network,* https://www.amazon.in/Make-Your-Own-Neural-Network-ebook/dp/B01EER4Z4G
- Nick Bostrom's *Superintelligence,* https://en.wikipedia.org/wiki/Superintelligence:_Paths,_Dangers,_Strategies
- Pedro Domingos's *The Master Algorithm,* https://en.wikipedia.org/wiki/The_Master_Algorithm
- *Deep Learning Book,* http://www.deeplearningbook.org/

2
Creating a Real-Estate Price Prediction Mobile App

In the previous chapter, we covered the theoretical fundamentals; this chapter, on the other hand, will cover the setup of all the tools and libraries.

First, we are going to set up our environment to build a Keras model to predict house prices with real estate data. Then we are going to serve this model using a RESTful API built using Flask. Next, we will set up our environment for Android and create an app that will consume this RESTful API to predict the house price based on features of real estate. Finally, we will repeat the same exercise for iOS.

The focus of this chapter is on the setup, tools, libraries, and exercising the concepts learned in Chapter 1, *Artificial Intelligence Concepts and Fundamentals*. The use case is designed to be simple, yet adaptable enough to accommodate similar use-cases. By the end of the chapter, you will be comfortable creating a mobile app for prediction or classification use cases.

In this chapter, we will cover the following topics:

- Setting up the artificial intelligence environment
- Building an ANN model for prediction using Keras and Tenserflow
- Serving the model as an API
- Creating an Android app to predict house prices
- Creating an iOS app to predict house prices

Setting up the artificial intelligence environment

The first thing to do is install Python. We are going to use Python throughout this book for all our **artificial intelligence** (**AI**) tasks. There are two ways to install Python, either through the downloadable executable file provided from `https://www.python.org/downloads/` or via Anaconda. Our approach will be the latter, that is, using Anaconda.

Downloading and installing Anaconda

Now, let's go to the official Anaconda installation page (`https://conda.io/docs/user-guide/install/index.html#regular-installation`) and choose the appropriate option based on your operating system:

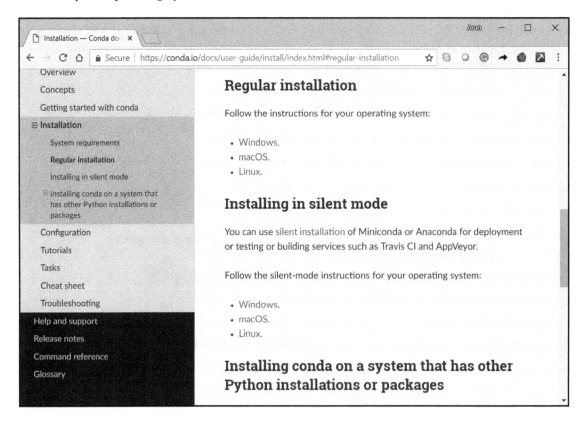

Follow the instructions as per the documentation. The installation takes some time.

Once it is installed, let's test the installation. Open the command prompt and type the
`conda list` command. You should see a list of libraries and packages that have been
installed as part of the Anaconda installation:

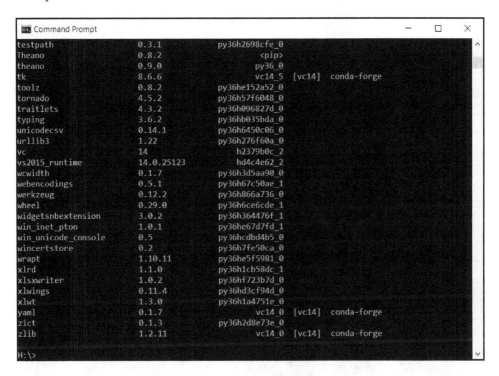

If you do not get this output, please follow the official documentation page we saw and try
again.

Advantages of Anaconda

Let's discuss a few positives of using a package-management tool:

- Anaconda lets us create environments to install libraries and packages. This
  environment is completely independent of the operating system or admin
  libraries. This means we can create user-level environments with custom versions
  of libraries for specific projects, which helps us port the project across operating
  systems with minimal effort.
- Anaconda can have multiple environments with different versions of Python and
  supporting libraries. This way, any version mismatch can be avoided and is not
  affected by existing packages and libraries of the operating system.

- Anaconda comes preloaded with most of necessary packages and libraries for data-science-related tasks, including a highly popular and interactive Python editor called Jupyter Notebook. Throughout this book, we will be using Jupyter Notebook a lot, mostly when we need to interactively code our tasks.

Creating an Anaconda environment

We will create an environment named `ai-projects` that uses Python version 3.6. All our dependencies are going to be installed in this environment:

```
conda create -n ai-projects python=3.6 anaconda
```

Now, proceed and accept the prompts that you are presented with, you should get an output that looks as follows:

```
Command Prompt                                              —    □    ×

  wheel:                        0.31.1-py36_0
  widgetsnbextension:           3.2.1-py36_0
  win_inet_pton:                1.0.1-py36he67d7fd_1
  win_unicode_console:          0.5-py36hcdbd4b5_0
  wincertstore:                 0.2-py36h7fe50ca_0
  winpty:                       0.4.3-4
  wrapt:                        1.10.11-py36he5f5981_0
  xlrd:                         1.1.0-py36h1cb58dc_1
  xlsxwriter:                   1.0.4-py36_0
  xlwings:                      0.11.8-py36_0
  xlwt:                         1.3.0-py36h1a4751e_0
  yaml:                         0.1.7-hc54c509_2
  zeromq:                       4.2.5-hc6251cf_0
  zict:                         0.1.3-py36h2d8e73e_0
  zlib:                         1.2.11-h8395fce_2

Proceed ([y]/n)? y

#
# To activate this environment, use:
# > activate ai-projects
#
# To deactivate an active environment, use:
# > deactivate
#
# * for power-users using bash, you must source
#

H:\>
```

Before we start installing the dependencies, we need to activate the environment we just created using the `activate ai-projects` command, or `source activate ai-projects` if you are using bash shell. The prompt will change to indicate that the environment has been activated:

```
Command Prompt                                    —    □    ×

#
# To activate this environment, use:
# > activate ai-projects
#
# To deactivate an active environment, use:
# > deactivate
#
# * for power-users using bash, you must source
#


H:\>activate ai-projects

(ai-projects) H:\>
```

Installing dependencies

First, let's install TensorFlow. It is an open source framework for building **Artificial Neural Network (ANN)**:

```
pip install tensorflow
```

You should see the following output, which indicates a successful installation:

We can also manually check the installation. Type `python` to open the Python prompt on the command line. Once inside the Python prompt, type `import tensorflow` and hit *Enter*. You should see the following output:

Type `exit()` to return to the default command line, keep in mind that we are still inside the `ai-projects` conda environment.

Next, we are going to install Keras, a wrapper over TensorFlow that makes designing deep neural networks much more intuitive. We continue to use the `pip` command:

```
pip install keras
```

On successful installation, we should see the following output:

To manually check the installation, type `python` to open the Python prompt on the command line. Once inside the Python prompt, type `import keras` and hit *Enter*. You should see the following output, with no errors. Observe that the output mentions that Keras is using TensorFlow as its backend:

Great! We have now installed the main dependencies required to create our very own neural networks. Let's go ahead and build an ANN to predict real estate prices.

Building an ANN model for prediction using Keras and TensorFlow

Now that we have our libraries installed, let's create a folder called `aibook` and within that create another folder called `chapter2`. Move all the code for this chapter into the `chapter2` folder. Make sure that the conda environment is still active (the prompt will start with the environment name):

Once within the `chapter2` folder, type `jupyter notebook`. This will open an interactive Python editor on the browser.

Use the **New** dropdown in the top-right corner to create a new **Python 3** notebook:

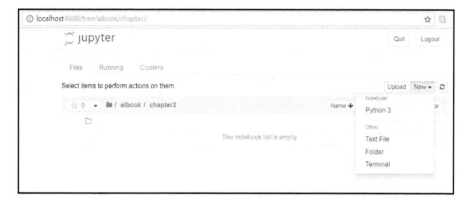

We are now ready to build our first ANN using Keras and TensorFlow, to predict real estate prices:

1. Import all the libraries that we need for this exercise. Use the first cell to import all the libraries and run it. Here are the four main libraries we will use:

   - pandas: We use this to read the data and store it in a dataframe
   - sklearn: We use this to standardize data and for k-fold cross-validation
   - keras: We use this to build our sequential neural network
   - numpy: We use numpy for all math and array operations

   Let's import these libraries:

   ```
   import numpy
   import pandas as pd
   from keras.models import Sequential
   from keras.layers import Dense
   from keras import optimizers
   from keras.wrappers.scikit_learn import KerasRegressor
   from sklearn.model_selection import cross_val_score
   from sklearn.model_selection import KFold
   from sklearn.preprocessing import StandardScaler
   from sklearn.pipeline import Pipeline
   ```

2. Load the real estate housing data using pandas:

   ```
   dataframe = pd.read_csv("housing.csv", sep=',', header=0)
   dataset = dataframe.values
   ```

3. To view the feature variables, the target variables, and a few rows of the data, enter the following:

```
dataframe.head()
```

This output will be a few rows of `dataframe`, which is shown in the following screenshot:

	BIZPROP	ROOMS	AGE	HIGHWAYS	TAX	PTRATIO	LSTAT	VALUE
0	19.58	7.489	90.8	5	403	14.7	1.73	50.0
1	19.58	7.802	98.2	5	403	14.7	1.92	50.0
2	19.58	8.375	93.9	5	403	14.7	3.32	50.0
3	19.58	7.929	96.2	5	403	14.7	3.70	50.0
4	2.46	7.831	53.6	3	193	17.8	4.45	50.0

The dataset has eight columns, details of each column are given as follows:

- **BIZPROP**: Proportion of non-retail business acres per town
- **ROOMS**: Average number of rooms per dwelling
- **AGE**: Proportion of owner-occupied units built before 1940
- **HIGHWAYS**: Index of accessibility to radial highways
- **TAX**: Full-value property tax rate per $10,000
- **PTRATIO**: Pupil-to-teacher ratio by town
- **LSTAT**: Percentage of lower status of the population
- **VALUE**: Median value of owner-occupied homes in thousand dollars (target variable)

In our use case, we need to predict the **VALUE** column, so we need to split the dataframe into features and target values. We will use a 70/30 split, that is, 70% of data for training and 30% data for testing:

```
features = dataset[:,0:7]
target = dataset[:,7]
```

Also, to make sure we can reproduce the results, let's set a seed for random generation. This random function is used during cross-validation to randomly sample the data:

```
# fix random seed for reproducibility
seed = 9
numpy.random.seed(seed)
```

Now we are ready to build our ANN:

1. Create a sequential neural network that has a simple and shallow architecture.

2. Make a function called `simple_shallow_seq_net()` that will define the architecture of the neural network:

```
def simple_shallow_seq_net():
   # create a sequential ANN
    model = Sequential()
    model.add(Dense(7, input_dim=7, kernel_initializer='normal',
activation='sigmoid'))
    model.add(Dense(1, kernel_initializer='normal'))
    sgd = optimizers.SGD(lr=0.01)
    model.compile(loss='mean_squared_error', optimizer=sgd)
    return model
```

3. The function does the following:

```
model = Sequential()
```

4. A sequential model is instantiated – a sequential model is an ANN model built using a linear stack of layers:

```
model.add(Dense(7, input_dim=7, kernel_initializer='normal',
activation='sigmoid'))
```

5. Here, we are adding a dense layer or fully-connected layer with seven neurons that are added to this sequential network. This layer accepts an input with 7 features (since there are seven input or features for predicting house price), which is indicated by the `input_dim` parameter. The weights of all the neurons in this layer are initialized using a random normal distribution, as indicated by the `kernel_initializer` parameter. Similarly, all the neurons of this layer use the sigmoid activation function, as indicated by the `activation` parameter:

```
model.add(Dense(1, kernel_initializer='normal'))
```

6. Add another layer with a single neuron initialized using a random normal distribution:

```
sgd = optimizers.SGD(lr=0.01)
```

7. Set the network to use **Scalar Gradient Descent** (**SGD**) to learn, usually specified as `optimizers`. We also indicate that the network will use a learning rate (`lr`) of `0.01` at every step of learning:

```
model.compile(loss='mean_squared_error', optimizer=sgd)
```

8. Indicate that the network needs to use the **mean squared error** (**MSE**) cost function to measure the magnitude of the error rate of the model, and use the SGD optimizer to learn from the wrongness measured or loss of the model:

```
return model
```

Finally, the function returns a model with the defined specifications.

The next step is to set a random seed for reproducibility; this random function is used to split the data into training and validation. The method used is k-fold validation, where the data is randomly divided into 10 subsets for training and validation:

```
seed = 9
kfold = KFold(n_splits=10, random_state=seed)
```

Now, we need to fit this model to predict a numerical value (house price, in this case), therefore we use KerasRegressor. KerasRegressor is a Keras wrapper used to access the regression estimators for the model from sklearn:

```
estimator = KerasRegressor(build_fn=simple_shallow_seq_net1, epochs=100,
batch_size=50, verbose=0)
```

Note the following:

- We pass simple_shallow_seq_net as a parameter to indicate the function that returns the model.
- The epochs parameter indicates that every sample needs to go through the network at least 100 times.
- The batch_size parameter indicates that during every learning cycle of the network there are 50 training samples used at a time.

The next step is to train and cross-validate across the subsets of the data and print the MSE, which is the measure of how well the model performs:

```
results = cross_val_score(estimator, features, target, cv=kfold)
print("simple_shallow_seq_model:(%.2f) MSE" % (results.std()))
```

This will output the MSE – as you can see, it is pretty high and we need to make this value as low as possible:

```
simple_shallow_seq_net:(163.41) MSE
```

Save this model for later use:

```
estimator.fit(features, target)
estimator.model.save('simple_shallow_seq_net.h5')
```

Great, we have built and saved our first neural net to predict real estate price. Our next efforts are to improve the neural net. The first thing to try before fiddling with the network parameters is to improve its performance (lower the MSE) when we standardize the data and use it:

```
estimators = []
estimators.append(('standardize', StandardScaler()))
estimators.append(('estimator',
KerasRegressor(build_fn=simple_shallow_seq_net, epochs=100, batch_size=50,
verbose=0)))
pipeline = Pipeline(estimators)
```

In the preceding code, we created a pipeline to standardize the data and then use it during every learning cycle of the network. In the following code block, we train and cross-evaluate the neural network:

```
results = cross_val_score(pipeline, features, target, cv=kfold)
print("simple_std_shallow_seq_net:(%.2f) MSE" % (results.std()))
```

This will output a much better MSE than before, hence standardizing and using the data makes a difference:

```
simple_std_shallow_seq_net:(65.55) MSE
```

Saving this model is slightly different than before as we have used `pipeline` to fit the model:

```
pipeline.fit(features, target)
pipeline.named_steps['estimator'].model.save('standardised_shallow_seq_net.
h5')
```

Let's now fiddle with our network to see whether we can get better results. We can start by creating a deeper network. We will increase the number of hidden or fully-connected layers and use both the `sigmoid` and `tanh` activation functions in alternate layers:

```
def deep_seq_net():
    # create a deep sequential model
    model = Sequential()
    model.add(Dense(7, input_dim=7, kernel_initializer='normal',
activation='sigmoid'))
    model.add(Dense(7,activation='tanh'))
    model.add(Dense(7,activation='sigmoid'))
    model.add(Dense(7,activation='tanh'))
    model.add(Dense(1, kernel_initializer='normal'))
    sgd = optimizers.SGD(lr=0.01)
    model.compile(loss='mean_squared_error', optimizer=sgd)
    return model
```

The next block of code is used to standardize the variables in the training data and then fit the shallow neural net model to the training data. Create the pipeline and fit the model using standardized data:

```
estimators = []
estimators.append(('standardize', StandardScaler()))
estimators.append(('estimator', KerasRegressor(build_fn=deep_seq_net,
epochs=100, batch_size=50, verbose=0)))
pipeline = Pipeline(estimators)
```

Now, we need to cross-validate the fit model across the subsets of the data and print the MSE:

```
results = cross_val_score(pipeline, features, target, cv=kfold)
print("simple_std_shallow_seq_net:(%.2f) MSE" % (results.std()))
```

This will output an MSE that is better than the previous shallow networks that we created:

```
deep_seq_net:(58.79) MSE
```

Save the model for later use:

```
pipeline.fit(features, target)
pipeline.named_steps['estimator'].model.save('deep_seq_net.h5')
```

So, we get better results when we increase the depth (layers) of the network. Now, let's see what happens when we widen the network, that is, increase the number of neurons (nodes) in each layer. Let's define a deep and wide network to tackle the problem, we increase the neurons in each layer to 21. Also, this time around, we will use the relu and sigmoid activation functions for the hidden layers:

```
def deep_and_wide_net():
    # create a sequential model
    model = Sequential()
    model.add(Dense(21, input_dim=7, kernel_initializer='normal',
activation='relu'))
    model.add(Dense(21,activation='relu'))
    model.add(Dense(21,activation='relu'))
    model.add(Dense(21,activation='sigmoid'))
    model.add(Dense(1, kernel_initializer='normal'))
    sgd = optimizers.SGD(lr=0.01)
    model.compile(loss='mean_squared_error', optimizer=sgd)
    return model
```

The next block of code is used to standardize the variables in the training data and then fit the deep and wide neural net model to the training data:

```
estimators = []
estimators.append(('standardize', StandardScaler()))
estimators.append(('estimator', KerasRegressor(build_fn=deep_and_wide_net,
epochs=100, batch_size=50, verbose=0)))
pipeline = Pipeline(estimators)
```

Now, we need to cross-validate the fit model across the subsets of the data and print the MSE:

```
results = cross_val_score(pipeline, features, target, cv=kfold)
print("deep_and_wide_model:(%.2f) MSE" % (results.std()))
```

This time, the MSE is again better than the previous networks we created. This is a good example of how a deeper network with more neurons abstracts the problem better:

```
deep_and_wide_net:(34.43) MSE
```

Finally, save the network for later use. The saved network model will be used in the next section and served within a REST API:

```
pipeline.fit(features, target)
pipeline.named_steps['estimator'].model.save('deep_and_wide_net.h5')
```

So far, we have been able to build a sequential neural network for prediction using various network architectures. As an exercise, try the following:

- Experiment with the shape of the network; play around with the depth and width of the network to see how it impacts the output
- Try out the various activation functions (https://keras.io/activations/)
- Try out the various initializers, here we have only used the random normal initializer (https://keras.io/initializers/)
- The data we used here is for demonstrating the technique, so try out different use cases for prediction using the preceding technique on other datasets (https://data.world/datasets/prediction)

We will learn more about optimizers and regularizers, which are other parameters you can use to tune the network, in Chapter 4, *Building a Machine Vision Mobile App to Classify Flower Species*. The complete code for our ANN model creation is available as a Python notebook named sequence_networks_for_prediction.ipynb.

Serving the model as an API

Now that we have created a model for prediction, the next thing is to serve this model via a RESTful API. To achieve this, we will use lightweight Python framework called Flask: `http://flask.pocoo.org/`.

Let's start by installing the `Flask` library in our conda environment if it does not already exist:

```
pip install Flask
```

Building a simple API to add two numbers

Now we will build a very simple API to get a grip on the `Flask` library and framework. This API will accept a JSON object with two numbers and return the sum of the numbers as a response.

Open a new notebook from your Jupyter home page:

1. Import all the libraries we need and create an app instance:

    ```
    from flask import Flask, request
    app = Flask(__name__)
    ```

2. Create the index page for the RESTful API using the `route()` decorator:

    ```
    @app.route('/')
    def hello_world():
     return 'This is the Index page'
    ```

3. Create a `POST` API to add two numbers using the `route()` decorator. This API accepts a JSON object with the numbers to be added:

    ```
    @app.route('/add', methods=['POST'])
    def add():
        req_data = request.get_json()
        number_1 = req_data['number_1']
        number_2 = req_data['number_2']
        return str(int(number_1)+int(number_2))
    ```

Save the Python notebook and use the **File** menu to download the notebook as a Python file. Place the Python file in the same directory as the model file.

Start a new command terminal and traverse to the folder with this Python file and the model. Make sure to activate the conda environment and run the following to start a server running the simple API:

- If you are using Windows, enter the following:

    set FLASK_APP=simple_api

- If you aren't using Windows, enter this:

    export FLASK_APP=simple_api

    Then type the following:

    flask run

    You should see the following output when the server starts:

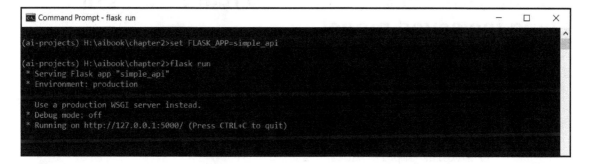

Open the browser and paste this address in the URL bar to go to the index page: `http://127.0.0.1:5000/`. Here is the output:

Next, we will use `curl` to access the `POST` API that adds two numbers. Open a new terminal and enter the following curl command to test the `/add` API. The numbers to add in this example are 1 and 2, and this is passed as a JSON object:

```
curl -i -X POST -H "Content-Type: application/json" -d
"{\"number_1\":\"1\",\"number_2\":\"2\"}" http://127.0.0.1:5000/add
```

We will get a response with the sum of the numbers if there are no errors:

```
Command Prompt                                                    —    □    ×

C:\>curl -i -X POST -H "Content-Type: application/json" -d "{\"number_1\":\"1\",\"number_2\":\"2\"}" http://127.0.0.1:50
00/add
HTTP/1.0 200 OK
Content-Type: text/html; charset=utf-8
Content-Length: 1
Server: Werkzeug/0.14.1 Python/3.6.5
Date: Mon, 23 Jul 2018 21:52:33 GMT

3
```

The complete code for the simple API is available as a Python notebook named
`simple_api.ipynb` and as a Python file named `simple_api.py`.

Building an API to predict the real estate price using the saved model

Now that we have seen how `Flask` works, we need to implement an API to serve the
model we built previously. Start a new Jupyter Notebook and follow these steps:

1. Import the required Python modules and create a Flask app instance:

```
from flask import Flask, request
from keras.models import load_model
from keras import backend as K

import numpy
app = Flask(__name__)
```

2. Create `Index` page for the RESTful API using the `route()` decorator:

```
@app.route('/')
def hello_world():
    return 'Index page'
```

3. Create a `POST` API to predict house price using the `route()` decorator. This
   accepts a JSON object with all the features required to predict the house or real
   estate price:

```
@app.route('/predict', methods=['POST'])
def add():
    req_data = request.get_json()
     bizprop = req_data['bizprop']
    rooms = req_data['rooms']
```

```
    age = req_data['age']
    highways = req_data['highways']
    tax = req_data['tax']
    ptratio = req_data['ptratio']
    lstat = req_data['lstat']
    # This is where we load the actual saved model into new
variable.
    deep_and_wide_net = load_model('deep_and_wide_net.h5')
    # Now we can use this to predict on new data
    value =
deep_and_wide_net.predict_on_batch(numpy.array([[bizprop, rooms,
age , highways , tax , ptratio , lstat]], dtype=float))
    K.clear_session()

    return str(value)
```

Save the Python notebook and use the **File** menu to download the notebook as a Python file. Place the Python file in the same directory as the model file.

Start a new command terminal and traverse to the folder with this Python file and the model. Make sure to activate the conda environment and run the following to start a server that runs the simple API:

- If you are using Windows, enter the following:

  set FLASK_APP=predict_api

- If you aren't using Windows, use this:

  export FLASK_APP= predict_api

  Then type the following:

  flask run

Next, we will use `curl` to access the POST API that predicts house prices. Open a new terminal and enter the following `curl` command to test the /predict API. We can pass the features to be used as input for the model as a JSON object:

```
curl -i -X POST -H "Content-Type: application/json" -d
"{\"bizprop\":\"1\",\"rooms\":\"2\",\"age\":\"1\",\"highways\":\"1\",\"tax\
":\"1\",\"ptratio\":\"1\",\"lstat\":\"1\"}" http://127.0.0.1:5000/predict
```

This will output the house price for the features provided using our prediction model:

```
Command Prompt                                              —    □    ×

C:\>curl -i -X POST -H "Content-Type: application/json" -d "{\"bizprop\":\"1\",\"rooms\":\"2\",\"age\":\"1\",\"highways\
":\"1\",\"tax\":\"1\",\"ptratio\":\"1\",\"lstat\":\"1\"}" http://127.0.0.1:5000/predict
HTTP/1.0 200 OK
Content-Type: text/html; charset=utf-8
Content-Length: 13
Server: Werkzeug/0.14.1 Python/3.6.5
Date: Mon, 23 Jul 2018 22:06:21 GMT

[[14.462999]]
```

That's it! We just built an API to serve our prediction model and tested it using `curl`. The complete code for the prediction API is available as a Python notebook named `predict_api.ipynb` and as a Python file named `simple_api.py`.

Next, we are going to see how to make a mobile app that will use the API that hosts our model. We will start by creating an Android app that uses the prediction API and then repeat the same task on an iOS app.

Creating an Android app to predict house prices

In this section, we are going to consume the model through the RESTful API via an Android app. The purpose of this section is to demonstrate how a model can be consumed and used by an Android app. Here, we have assumed that you are familiar with the basics of Java programming. The same approach can be used for any similar use case, even on web apps. The following steps are covered in this section:

- Downloading and installing Android Studio
- Creating a new Android project with a single screen
- Designing the layout of the screen
- Adding a functionality to accept input
- Adding a functionality to consume the RESTful API that serves the model
- Additional notes

Downloading and installing Android Studio

Android Studio is the development environment and sandbox for Android app development. All our Android projects will be made using Android Studio. We can use Android Studio to create, design, and test our apps before publishing them.

Head over to the official Android Studio download page, `https://developer.android.com/studio/`, and choose the version that matches your OS. In this case, we are using a Windows executable:

Run the executable once it is downloaded to start the installation process. You will be presented with progressive installation menu choices. Choose **Next** and progress through the installation process. Most of the options are chosen as default during the installation steps.

Creating a new Android project with a single screen

Now that we have installed Android Studio, we will create a simple app to estimate the price of real estate based on certain input.

Once we start Android Studio, it gives us a menu to start creating projects. Click on the **Start a new Android Studio project** option:

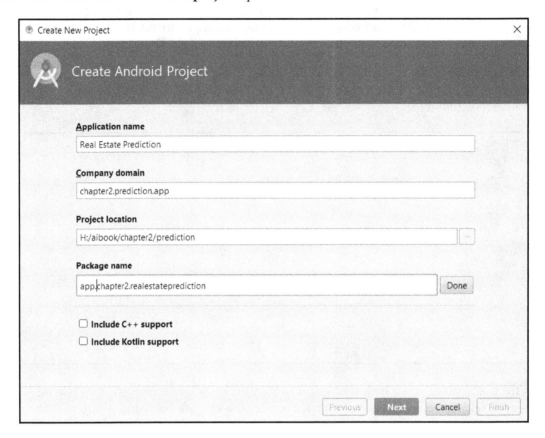

The next dialog is to select the **Application name** and **Project location**. Choose whatever you want and click **Next**:

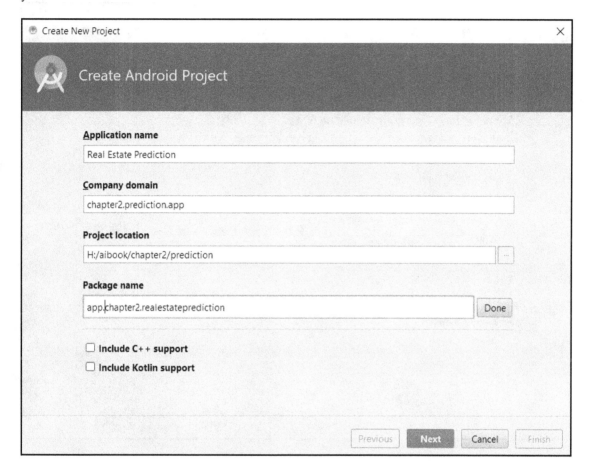

Next, choose the target versions for the application to run on:

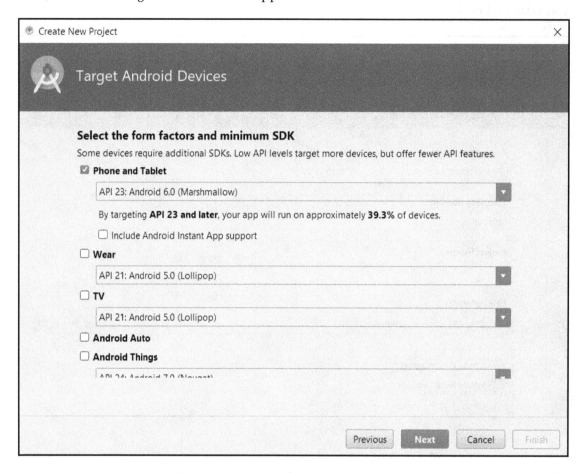

Then choose a screen for the app; in this case, select an **Empty Activity**:

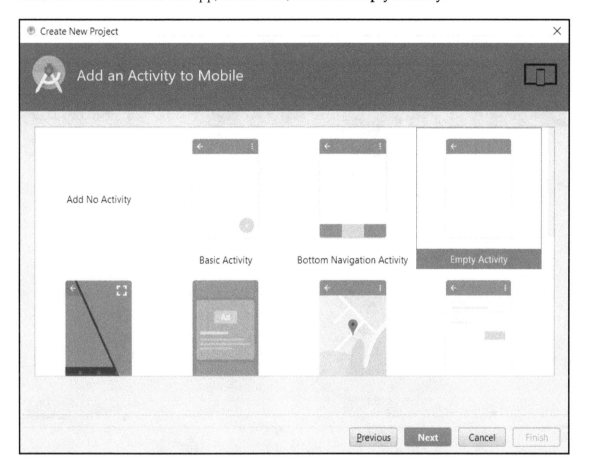

Choose the screen or **Activity Name** and the corresponding name for the layout or design of the activity screen:

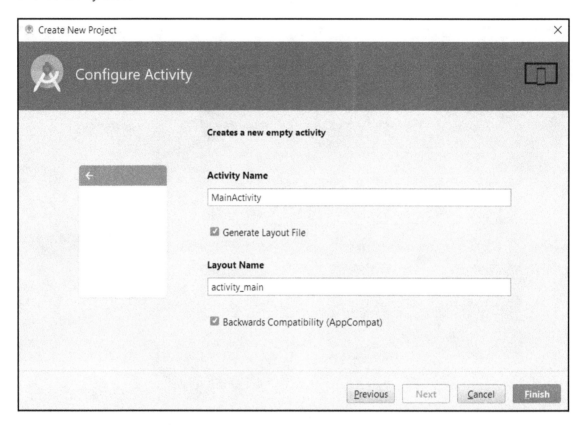

The project should load in a few seconds after the build is complete. In the project structure, there are three main folders:

- **manifests**: This folder contains the manifest file used for permissions and application versioning.
- **java**: This folder has all the Java code files (**java | app | chapter2 | realestateprediction | MainActivity.java**).
- **res**: This folder has all the layout files and media files used in the application (**res | layout | activity_main.xml**):

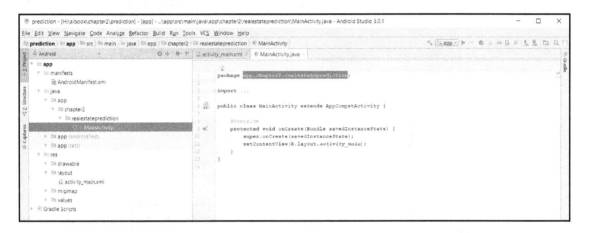

Designing the layout of the screen

Let's design the screen that will accept the factors of the model we created as input. The screen will have seven input boxes to accept the factors, one button, and an output textbox to display the predicted result:

Traverse to the **layout** folder in **res** and select the **activity_layout.xml** file to open in the editing panel. Choose the **Text** option at the bottom to view the existing XML for the layout:

```
activity_main.xml      MainActivity.java

1    <?xml version="1.0" encoding="utf-8"?>
2    <android.support.constraint.ConstraintLayout xmlns:android="http://schemas.android.com/apk/res/android"
3        xmlns:app="http://schemas.android.com/apk/res-auto"
4        xmlns:tools="http://schemas.android.com/tools"
5        android:layout_width="match_parent"
6        android:layout_height="match_parent"
7        tools:context="app.chapter2.realestateprediction.MainActivity">
8
9        <TextView
10           android:layout_width="wrap_content"
11           android:layout_height="wrap_content"
12           android:text="Hello World!"
13           app:layout_constraintBottom_toBottomOf="parent"
14           app:layout_constraintLeft_toLeftOf="parent"
15           app:layout_constraintRight_toRightOf="parent"
16           app:layout_constraintTop_toTopOf="parent" />
17
18   </android.support.constraint.ConstraintLayout>
19

Design   Text
```

Now, replace the existing XML code with the new design template for the app. Please refer to the **activity_layout.xml** code file in the **Android** folder for the full design template. The following is only a skeletal reference of the XML code template:

```xml
<?xml version="1.0" encoding="utf-8"?>
<ScrollView xmlns:android="http://schemas.android.com/apk/res/android"
    android:layout_width="match_parent"
    android:layout_height="match_parent"
    android:fillViewport="true">
    <RelativeLayout
        android:layout_width="match_parent"
        android:layout_height="match_parent"
        >
        <TextView
            android:id="@+id/bizprop"/>
        <EditText
            android:id="@+id/bizprop-edit"/>
        <TextView
            android:id="@+id/rooms"/>
```

```
        <EditText
            android:id="@+id/rooms-edit"/>
        <TextView
            android:id="@+id/age"/>
        <EditText
            android:id="@+id/age-edit"/>
        <TextView
            android:id="@+id/highways"/>
        <EditText
            android:id="@+id/highways-edit"/>
        <TextView
            android:id="@+id/tax"/>
        <EditText
            android:id="@+id/tax-edit"/>
        <TextView
            android:id="@+id/ptratio"/>
        <EditText
            android:id="@+id/ptratio-edit"/>
        <TextView
            android:id="@+id/lstat"/>
        <EditText
            android:id="@+id/lstat-edit"/>
        <Button
            android:id="@+id/button"/>
        <TextView
            android:id="@+id/value"/>
    </RelativeLayout>
</ScrollView>
```

Here, we have designed a layout to accept the seven factors as input, as follows:

- **BIZPROP**: Proportion of non-retail business acres per town
- **ROOMS**: Average number of rooms per dwelling
- **AGE**: Proportion of owner-occupied units built prior to 1940
- **HIGHWAYS**: Index of accessibility to radial highways
- **TAX**: Full-value property-tax rate per $10,000

- **PTRATIO**: Pupil-to-teacher ratio by town
- **LSTAT**: Percentage of lower status of the population

There is also a button and a textbox to display the output. The predicted value is displayed when the button is clicked.

To view the design of the activity, run the **Run app** option in the **run** menu from the top menu bar. The first time you run it, the environment will ask you to create a virtual device to test your app. You can either create an **Android Virtual Device** (**AVD**) or use the traditional method, that is, use an USB cable to connect your Android phone to the PC so that you can run the output directly on your device:

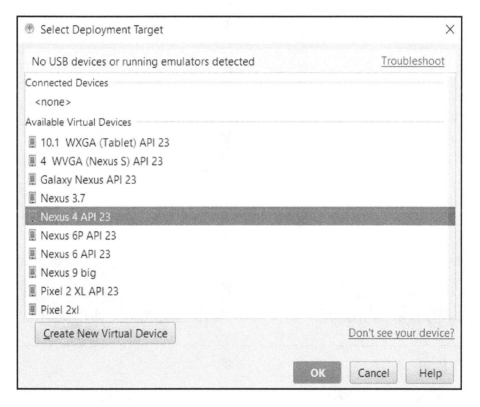

You should see the design of the scrollable layout once the app starts on the device or AVD emulator:

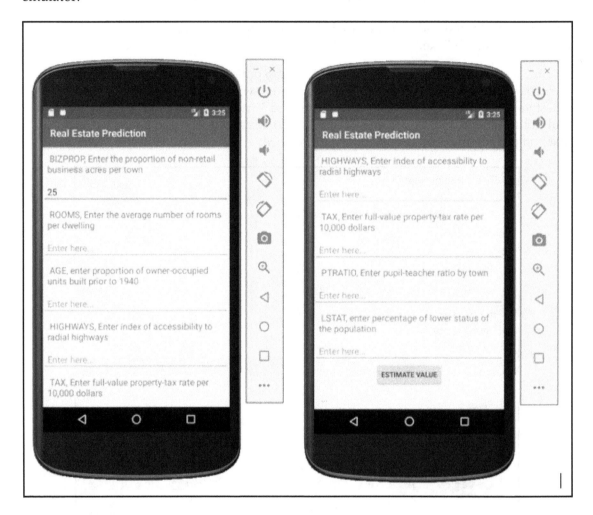

Adding a functionality to accept input

Now we need to accept the input and create a map to hold the values. We will then convert the map into a JSON object so that it can be passed as the data in the POST API request.

Traverse to the `MainActivity.java` file and open it in the edit panel of Android Studio. Declare the following class variables:

```java
private EditText bizprop, rooms, age, highways, tax, ptratio, lstat;
private Button estimate;
private TextView value;
```

You will find a function called `onCreate()` that is already created. Add the following code into the `onCreate()` function to initialize the elements of the layout:

```java
bizprop = (EditText) findViewById(R.id.bizprop_edit);
rooms = (EditText) findViewById(R.id.rooms_edit);
age = (EditText) findViewById(R.id.age_edit);
highways = (EditText) findViewById(R.id.highways_edit);
tax = (EditText) findViewById(R.id.tax_edit);
ptratio = (EditText) findViewById(R.id.ptratio_edit);
lstat = (EditText) findViewById(R.id.lstat_edit);
value = (TextView) findViewById(R.id.value);
estimate = (Button) findViewById(R.id.button);
```

Now add another function called `makeJSON()` to the Java class. This function accepts the values from the edit boxes and returns the JSON object we need to pass as data in our API call:

```java
public JSONObject makeJSON() {

    JSONObject jObj = new JSONObject();
    try {

        jObj.put("bizprop", bizprop.getText().toString());
        jObj.put("rooms", rooms.getText().toString());
        jObj.put("age", age.getText().toString());
        jObj.put("tax", tax.getText().toString() );
        jObj.put("highways", highways.getText().toString());
        jObj.put("ptratio", ptratio.getText().toString());
        jObj.put("lstat", lstat.getText().toString());
    } catch (Exception e) {
        System.out.println("Error:" + e);
    }

    Log.i("", jObj.toString());

    return jObj;
}
```

Adding a functionality to consume the RESTful API that serves the model

Now we need to hit the API with the data on a click of the button. To achieve this, we need the flowing helper functions:

- `ByPostMethod`: Accepts the URL as a `String` and returns the response as an `InputStream`. This function takes the server URL string that we created using the Flask framework and returns the response from the server as an input stream:

```
InputStream ByPostMethod(String ServerURL) {

    InputStream DataInputStream = null;
    try {
        URL url = new URL(ServerURL);

        HttpURLConnection connection = (HttpURLConnection)
        url.openConnection();
        connection.setDoOutput(true);
        connection.setDoInput(true);
        connection.setInstanceFollowRedirects(false);
        connection.setRequestMethod("POST");
        connection.setRequestProperty("Content-Type",
"application/json");
        connection.setRequestProperty("charset", "utf-8");
        connection.setUseCaches (false);
        DataOutputStream dos = new
DataOutputStream(connection.getOutputStream());
        dos.writeBytes(makeJSON().toString());
        //flushes data output stream.
        dos.flush();
        dos.close();
        //Getting HTTP response code
        int response = connection.getResponseCode();
        //if response code is 200 / OK then read Inputstream
        //HttpURLConnection.HTTP_OK is equal to 200
        if(response == HttpURLConnection.HTTP_OK) {
            DataInputStream = connection.getInputStream();
        }

    } catch (Exception e) {
        Log.e("ERROR CAUGHT", "Error in GetData", e);
    }
    return DataInputStream;

}
```

- ConvertStreamToString: This function accepts InputStream and returns a String of the response. The input stream returned from the previous function is processed as a string object by the following function:

```
String ConvertStreamToString(InputStream stream) {

    InputStreamReader isr = new InputStreamReader(stream);
    BufferedReader reader = new BufferedReader(isr);
    StringBuilder response = new StringBuilder();

    String line = null;
    try {

        while ((line = reader.readLine()) != null) {
            response.append(line);
        }

    } catch (IOException e) {
        Log.e("ERROR CAUGHT", "Error in ConvertStreamToString",
e);
    } catch (Exception e) {
        Log.e("ERROR CAUGHT", "Error in ConvertStreamToString",
e);
    } finally {

        try {
            stream.close();

        } catch (IOException e) {
            Log.e("ERROR CAUGHT", "Error in
ConvertStreamToString", e);

        } catch (Exception e) {
            Log.e("ERROR CAUGHT", "Error in
ConvertStreamToString", e);
        }
    }
    return response.toString();
}
```

- DisplayMessage: This functions updates the textbox with the response, which is the predicted value:

```
public void DisplayMessage(String a)
{
    value.setText(a);
}
```

A point to note is that whenever network calls are made on Android, it is a best practice to do it on a separate thread so it does not block the main **user interface** (**UI**) thread. So, we will write an inner class called `MakeNetworkCall` to achieve this:

```java
private class MakeNetworkCall extends AsyncTask<String, Void, String> {

    @Override
    protected void onPreExecute() {
        super.onPreExecute();
        DisplayMessage("Please Wait ...");
    }

    @Override
    protected String doInBackground(String... arg) {

        InputStream is = null;
        String URL = "http://10.0.2.2:5000/predict";
        Log.d("ERROR CAUGHT", "URL: " + URL);
        String res = "";

        is = ByPostMethod(URL);

        if (is != null) {
            res = ConvertStreamToString(is);
        } else {
            res = "Something went wrong";
        }
        return res;
    }

    protected void onPostExecute(String result) {
        super.onPostExecute(result);

        DisplayMessage(result);
        Log.d("COMPLETED", "Result: " + result);
    }
}
```

Please note that we used `http://10.0.2.2:5000/predict` instead of `http://127.0.0.1:5000/predict`. This is done because in Android, when we use the emulator, it accesses the local host via `10.0.2.2` instead of `127.0.0.1`. As the example is run on an emulator, we used `10.0.2.2`.

Finally, we need to add the functionality to call the API on click of the button. So, within the `oncreate()` method, insert the following code after the button has been initialized. This will initiate a background thread to access the API on a click of the button:

```
estimate.setOnClickListener(new View.OnClickListener() {
    @Override
    public void onClick(View v) {
        Log.i("CHECK", "CLICKED");

        new MakeNetworkCall().execute("http://10.0.2.2:5000/predict",
"Post");
    }
});
```

We need to add permissions to use internet within the `AndroidManifest.xml` file. Place the following code inside the `<manifest>` tag:

```
<uses-permission android:name="android.permission.INTERNET"></uses-permission>
```

Don't forget to run your Flask app. If you haven't already, make sure to run it within the conda environment activated:

```
set FLASK_APP=predict_api

flask run
```

That is all the code required to run and test the app on Android. Now run the app in the emulator, enter the details on the screen, hit the **ESTIMATE VALUE** button, and you'll get immediate results:

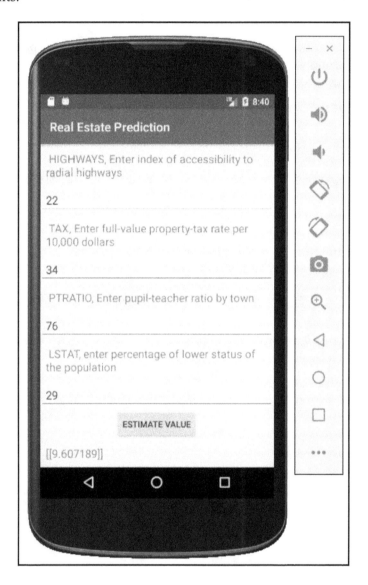

Additional notes

This was a demonstration of how we can use the AI models built on an Android device. Having said that, there are a lot of additional tasks that can be added to the existing app:

- Improve the UI design
- Add input checks to validate the data entered
- Host the Flask app (Heroku, AWS, and more)
- Publish the app

All these tasks are independent of our core AI theme, so can be addressed as an exercise for the reader.

Creating an iOS app to predict house prices

In this section, we are going to consume the model through the RESTful API via an iOS app. The purpose of this section is to demonstrate how a model can be consumed and used by an iOS app. Here, we have assumed that you are familiar with Swift programming. The same approach can be used for any similar use case. These are the following steps covered in this section:

- Downloading and installing Xcode
- Creating a new iOS project with a single screen
- Designing the layout of the screen
- Adding a functionality to accept input
- Adding a functionality to consume the RESTful API that serves the model
- Additional notes

Downloading and installing Xcode

You need a Mac (macOS 10.11.5 or later) to develop the iOS apps that are implemented in this book. Also, the latest version of Xcode is required to run those codes since it contains all the features that are necessary to design, develop, and debug any app.

To download the latest version of Xcode, follow these steps:

1. Open the App Store on your Mac (it's in the Dock by default).
2. Type Xcode in the search field, which is at the top-right corner. Then press the return key.

3. The first search result that turns up is the Xcode app.
4. Click on **Get** and then click on **Install App**.
5. Enter your Apple ID and password.
6. Xcode will be downloaded in your /Applications directory.

Creating a new iOS project with a single screen

Xcode includes several built-in app templates. We will start with a basic template: Single View Application. Open Xcode from the /Applications directory to create a new project.

If you are launching Xcode for the first time, it may ask you to agree to all the user agreements. Proceed by clicking on these prompts until Xcode is set up and ready to launch on your system.

Once we launch Xcode, the following window appears:

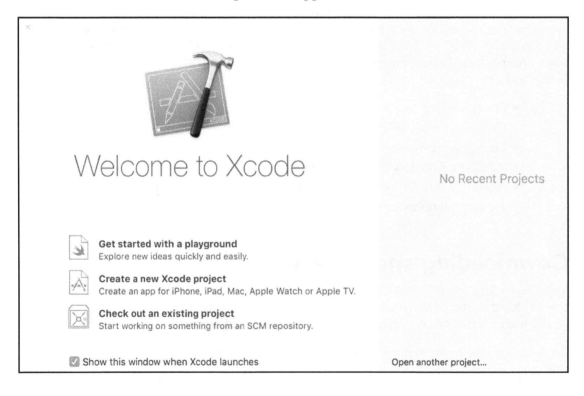

Click on **Create a new Xcode project**. A new window will open that displays a dialog box that allows us to select the required template:

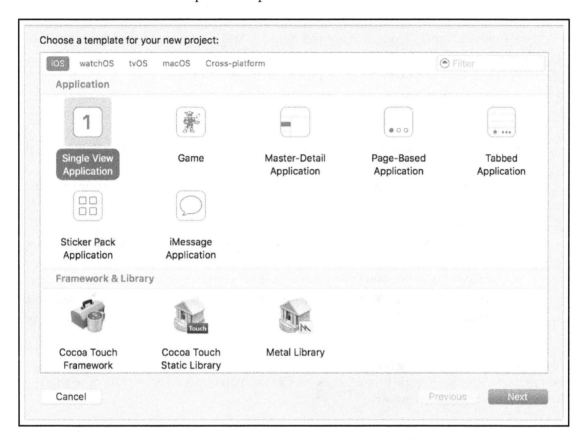

After we select a template, a dialog box appears. Here, you need to name your app for which you can use the following values. You can also choose some additional options for your project:

- **Product Name**: Xcode will use the product name that you entered to name both the project and the app.
- **Team**: If there's no value filled in, set the team to **None**.
- **Organization Name**: This is an optional field. You can either enter your organization's name or your name. You may also choose to leave this option blank.
- **Organization Identifier**: If you have an organization identifier, use that value. If you don't, use `com.example`.
- **Bundle Identifier**: This value is automatically generated based on your product name and organization identifier.
- **Language**: **Swift**.
- **Devices**: **Universal**. An app that runs on both iPhones and iPads is considered a universal app.
- **Use Core Data**: We don't need any core date. Hence, it remains unselected.
- **Include Unit Tests**: We need to include unit tests. So this option will be selected.
- **Include UI Tests**: We don't need to include any UI tests. Hence, this option stays unselected.

Now, click on **Next**. A dialog box will appear, where you need to select a location to save your project. After saving the project, click on **Create**. You new project will be opened by Xcode in the workspace window.

Designing the layout of the screen

Let's design the screen that will accept the factors of the model we created as input. The screen will have seven input boxes to accept the factors, one button, and an output textbox to display the predicted result:

Estimate the value of the real estate

Enter Real estate details:

BIZPROP, the proportion of non-retail business acres per town

ROOMS, the average number of rooms per dwelling

AGE, proportion of owner-occupied units built prior to 1940

HIGHWAYS, index of accessibility to radial highways

TAX, full-value property-tax rate per 10,000 dollars

PTRATIO, pupil-teacher ratio by town

LSTAT, Percentage of lower status of the population

Estimate

Let's work on the storyboard that is required for the app. What is a storyboard? A storyboard displays the screen of content and the transitions between that content. It gives us a visual representation of the application's UI. We get a **WYSIWYG** (short for **what you see is what you get**) editor where we can see the changes in real-time.

To open the storyboard, select the **Main.storyboard** option in the project navigator. This will open a canvas where we can design the screen. We can now add elements and design the canvas:

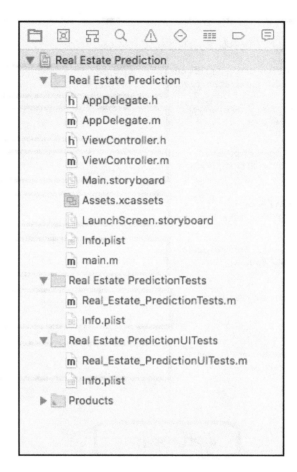

The same can also be coded instead of using the drag-and-drop approach. To do so, start by defining the text fields that are used as input in the ViewController class:

```
@interface ViewController ()<UITextFieldDelegate>
{
 UITextField*
bizropfeild, *roomsfeild, *agefeild, *highwaysfeild, *taxfeild, *ptratiofeild, *l
statfeild;
}
@end
```

Then, within the `CreateView` method, we implement the design of each text field. The following is the sample for the first couple of text fields; the same approach can be used for the rest of the text fields. The completed project code is available in the `chapter2_ios_prediction` folder.

First, create a header text field, `Estimate the value of real estate`:

```
UILabel *headerLabel = [[UILabel alloc]initWithFrame:CGRectMake(10, 20,
self.view.frame.size.width-20, 25)];
 headerLabel.font = [UIFont fontWithName:@"SnellRoundhand-Black" size:20];
//custom font
 headerLabel.backgroundColor = [UIColor clearColor];
 headerLabel.textColor = [UIColor blackColor];
 headerLabel.textAlignment = NSTextAlignmentCenter;
 headerLabel.text=@"Estimate the value of real estate";
 [self.view addSubview:headerLabel];


 UIView *sepratorLine =[[UIView alloc]initWithFrame:CGRectMake(0, 50,
self.view.frame.size.width, 5)];
 sepratorLine.backgroundColor=[UIColor blackColor];
 [self.view addSubview:sepratorLine];
```

Next, create another text field, `Enter real estate details`:

```
UILabel *detailLabel = [[UILabel alloc]initWithFrame:CGRectMake(10, 55,
self.view.frame.size.width-20, 25)];
 detailLabel.font = [UIFont fontWithName:@"SnellRoundhand-Black" size:18];
//custom font
 detailLabel.backgroundColor = [UIColor clearColor];
 detailLabel.textColor = [UIColor blackColor];
 detailLabel.textAlignment = NSTextAlignmentLeft;
 detailLabel.text=@"Enter real estate details";
 [self.view addSubview:detailLabel];
```

Next, create a field to enter the input for a proportion of non-retail business in acres:

```
UILabel *bizropLabel = [[UILabel alloc]initWithFrame:CGRectMake(5, 85,
self.view.frame.size.width-150, 35)];
 bizropLabel.font = [UIFont fontWithName:@"TimesNewRomanPSMT" size:12];
//custom font
 bizropLabel.backgroundColor = [UIColor clearColor];
 bizropLabel.numberOfLines=2;
 bizropLabel.textColor = [UIColor blackColor];
 bizropLabel.textAlignment = NSTextAlignmentLeft;
 bizropLabel.text=@"Bizrope, The proportion of non-retail business acres
per town";
 [self.view addSubview:bizropLabel];
```

```objectivec
bizropfeild = [[UITextField alloc]
initWithFrame:CGRectMake(self.view.frame.size.width-140, 85, 130, 35)];
 bizropfeild.delegate=self;
 bizropfeild.layer.borderColor=[UIColor blackColor].CGColor;
 bizropfeild.layer.borderWidth=1.0;
 [self.view addSubview:bizropfeild];
```

Now create a field to enter the input for the average number of rooms per dwelling:

```objectivec
UILabel *roomsLabel = [[UILabel alloc]initWithFrame:CGRectMake(5, 125,
self.view.frame.size.width-150, 35)];
 roomsLabel.font = [UIFont fontWithName:@"TimesNewRomanPSMT" size:12];
//custom font
 roomsLabel.backgroundColor = [UIColor clearColor];
 roomsLabel.numberOfLines=2;
 roomsLabel.textColor = [UIColor blackColor];
 roomsLabel.textAlignment = NSTextAlignmentLeft;
 roomsLabel.text=@"ROOMS, the average number of rooms per dwelling";
 [self.view addSubview:roomsLabel];

 roomsfeild = [[UITextField alloc]
initWithFrame:CGRectMake(self.view.frame.size.width-140, 125, 130, 35)];
 roomsfeild.delegate=self;
 roomsfeild.layer.borderColor=[UIColor blackColor].CGColor;
 roomsfeild.layer.borderWidth=1.0;
 [self.view addSubview:roomsfeild];
```

Then, create a button to hit the API:

```objectivec
UIButton *estimateButton = [UIButton
buttonWithType:UIButtonTypeRoundedRect];
 [estimateButton addTarget:self action:@selector(estimateAction)
 forControlEvents:UIControlEventTouchUpInside];
 estimateButton.layer.borderColor=[UIColor blackColor].CGColor;
 estimateButton.layer.borderWidth=1.0;
 [estimateButton setTitle:@"Estimate" forState:UIControlStateNormal];
 [estimateButton setTitleColor:[UIColor blackColor]
forState:UIControlStateNormal];
 estimateButton.frame = CGRectMake(self.view.frame.size.width/2-80, 375,
160.0, 40.0);
 [self.view addSubview:estimateButton];
```

Adding a functionality to accept input

Here, all the input from the text fields are packaged in an `NSString` object, which is used in the `POST` body of the request:

```
NSString *userUpdate =[NSString
stringWithFormat:@"bizprop=%@&rooms=%@&age=%@&highways=%@&tax=%@&ptratio=%@
&lstat=%@",bizropfeild.text,roomsfeild.text,agefeild.text,highwaysfeild.tex
t,taxfeild.text,ptratiofeild.text,lstatfeild.text];
```

Adding a functionality to consume the RESTful API that serves the model

Now we need to use the `NSURLSession` object to hit the RESTful API with the input from the activity screen:

```
//create the Method "GET" or "POST"
 [urlRequest setHTTPMethod:@"POST"];
 //Convert the String to Data
 NSData *data1 = [userUpdate dataUsingEncoding:NSUTF8StringEncoding];
 //Apply the data to the body
 [urlRequest setHTTPBody:data1];
 NSURLSession *session = [NSURLSession sharedSession];
 NSURLSessionDataTask *dataTask = [session dataTaskWithRequest:urlRequest
completionHandler:^(NSData *data, NSURLResponse *response, NSError *error)
{ }
```

Finally, display the output from the response received from the API:

```
NSHTTPURLResponse *httpResponse = (NSHTTPURLResponse *)response;
 if(httpResponse.statusCode == 200)
 {
 NSError *parseError = nil;
 NSDictionary *responseDictionary = [NSJSONSerialization
JSONObjectWithData:data options:0 error:&parseError];
 NSLog(@"The response is - %@",responseDictionary);
UILabel *outputLabel = [[UILabel alloc]initWithFrame:CGRectMake(5, 325,
self.view.frame.size.width-150, 35)];
 outputLabel.font = [UIFont fontWithName:@"TimesNewRomanPSMT" size:12];
//custom font
 outputLabel.backgroundColor = [UIColor clearColor];
 outputLabel.numberOfLines=2;
 outputLabel.textColor = [UIColor blackColor];
 outputLabel.textAlignment = NSTextAlignmentLeft;
 outputLabel.text = [responseDictionary valueForKey:@""];
```

```
[self.view addSubview:outputLabel];
}
```

The app can now run and is ready to be tested on the simulator.

Additional notes

This was a demonstration of how we can use AI models on an iOS device. Having said that, there are a lot of tasks that can be added to the existing app:

- Improve the UI design
- Add input checks to validate the data entered
- Host the Flask app (Heroku, AWS, and more)
- Publish the app

All these tasks are independent of our core AI theme, so they can be addressed as an exercise for the reader. The complete code and project files for both the Android and iOS apps are available as `chapter2_android_prediction` and `chapter2_ios_prediction`.

Summary

In this chapter, we explored the basic sequential network and consumed it on mobile devices. In the next chapter, we will take a look at a special kind of network called **Convolutional Neural Networks (CNN)**. CNNs are the most common networks used with Machine Vision. Our goal in the next chapter is to get comfortable with machine vision and to build our own custom-purpose CNNs.

3
Implementing Deep Net Architectures to Recognize Handwritten Digits

In the previous chapters, we have been through the essential concepts and have set up tools that are required for us to get our journey into **Artificial Intelligence** (**AI**) started. We also built a small prediction app to get our feet wet with the tools we will be using.

In this chapter, we are going to cover a more interesting and popular application of AI – Computer Vision, or Machine Vision. We will start by continuing from the previous chapter and ease into building **convolutional neural networks** (**CNN**), the most popular neural network type for Computer Vision. This chapter will also cover the essential concepts that were promised in Chapter 1, *Artificial Intelligence Concepts and Fundamentals*, but, in contrast, this chapter will have a very hands-on approach.

We will be covering the following topics in the chapter:

- Building a feedforward neural network to recognize handwritten digits
- Remaining concepts of neural networks
- Building a deeper neural network
- Introduction to computer vision

Building a feedforward neural network to recognize handwritten digits, version one

In this section, we will use the knowledge that we gained from the last two chapters to tackle a problem that has unstructured data – image classification. The idea is to take a dive into solving a Computer Vision task with the current setup and the basics of neural networks that we are familiar with. We have seen that feedforward neural networks can be used for prediction using structured data; let's try that on images to classify handwritten digits.

To solve this task, we are going to leverage the **MNSIT** database and use the handwritten digits dataset. MNSIT stands for **Modified National Institute of Standards and Technology**. It is a large database that's commonly used for training, testing, and benchmarking image-related tasks in Computer Vision.

The MNSIT digits dataset contains 60,000 images of handwritten digits, which are used for training the model, and 10,000 images of handwritten digits, which are used for testing the model.

From here out, we will be using Jupyter Notebook to understand and execute this task. So, please start your Jupyter Notebook and create a new Python Notebook if you have not already done so.

Once you have your notebook ready, the first thing to do, as always, is to import all the necessary modules for the task at hand:

1. Import numpy and set the seed for reproducibility:

   ```
   import numpy as np
   np.random.seed(42)
   ```

2. Load the Keras dependencies and the built-in MNSIT digits dataset:

   ```
   import keras
   from keras.datasets import mnist
   from keras.models import Sequential
   from keras.layers import Dense
   from keras.optimizers import SGD
   ```

3. Load the data into the training and test sets, respectively:

   ```
   (X_train, y_train), (X_test, y_test)= mnist.load_data()
   ```

4. Check the number of training images, along with the size of each image. In this case, the size of each image is 28 x 28 pixels:

```
X_train.shape
(60000, 28, 28)
```

5. Check the dependent variable, in this case, 60,000 cases with the right label:

```
y_train.shape
(60000,)
```

6. Check the labels for the first 100 training samples:

```
y_train [0 :99]
array([5, 0, 4, 1, 9, 2, 1, 3, 1, 4, 3, 5, 3, 6, 1, 7, 2, 8, 6, 9,
4, 0, 9,
       1, 1, 2, 4, 3, 2, 7, 3, 8, 6, 9, 0, 5, 6, 0, 7, 6, 1, 8, 7,
9, 3, 9,
       8, 5, 9, 3, 3, 0, 7, 4, 9, 8, 0, 9, 4, 1, 4, 4, 6, 0, 4, 5,
6, 1, 0,
       0, 1, 7, 1, 6, 3, 0, 2, 1, 1, 7, 9, 0, 2, 6, 7, 8, 3, 9, 0,
4, 6, 7,
       4, 6, 8, 0, 7, 8, 3], dtype=uint8)
```

7. Check the number of test images, along with the size of each image. In this case, the size of each image is 28 x 28 pixels:

```
X_test.shape
(10000, 28, 28)
```

8. Check the samples in the test data, which are basically 2D arrays of size 28 x 28:

```
X_test[0]
array([[  0,   0,   0,   0,   0,   0,   0,   0,   0,   0,   0,   0,
0,
          0,   0,   0,   0,   0,   0,   0,   0,   0,   0,   0,   0,
0,
          .
          .
       ,
          0,   0],
       [  0,   0,   0,   0,   0,   0,   0,   0,   0,   0, 121, 254,
207,
         18,   0,   0,   0,   0,   0,   0,   0,   0,   0,   0,   0,
0,
          0,   0,   0,   0,   0,   0,   0,   0,   0,   0,   0,   0,
0,
          0,   0]], dtype=uint8)
```

9. Check the dependent variable, in this case, 10,000 cases with the right label:

```
y_test.shape
(10000,)
```

10. The right label for the previous first sample in the test set is as follows:

```
y_test[0]
7
```

11. Now, we need to pre-process the data by converting it from a 28 x 28 2D array into a normalized 1D array of 784 elements:

```
X_train = X_train.reshape(60000, 784).astype('float32')
X_test = X_test.reshape(10000, 784).astype('float32')
X_train/=255
X_test /=255
```

12. Check the first sample of the pre-processed dataset:

```
X_test[0]
array([ 0.        ,  0.        ,  0.        ,  0.        ,  0.
,
        0.        ,  0.        ,  0.        ,  0.        ,  0.
,
        .
        .
        .
        0.        ,  0.        ,  0.        ,  0.        ,  0.
,
        0.47450981,  0.99607843,  0.99607843,  0.85882354,
0.15686275,
        0.        ,  0.        ,  0.        ,  0.        ,  0.
,
        0.        ,  0.        ,  0.        ,  0.        ,  0.
,
        0.        ,  0.        ,  0.        ,  0.        ,  0.
,
        0.        ,  0.        ,  0.        ,  0.        ,  0.
,
        0.        ,  0.        ,  0.        ,  0.47450981,
0.99607843,
        0.81176472,  0.07058824,  0.        ,  0.        ,  0.
,
        0.        ,  0.        ,  0.        ,  0.        ,  0.
,
        0.        ,  0.        ,  0.        ,  0.        ,  0.
,
```

```
    0.          ,  0.        ,  0.        ,  0.        ,  0.
 ,
    0.          ,  0.        ,  0.        ,  0.        ,  0.
 ,
    0.          ,  0.        ,  0.        ,  0.        ],
dtype=float32)
```

13. The next step is to one-hot code the labels; in other words, we need to convert the data type of the labels (zero to nine) from numeric into categorical:

```
n_classes=10
y_train= keras.utils.to_categorical(y_train ,n_classes)
y_test= keras.utils.to_categorical(y_test,n_classes)
```

14. View the first sample of the label that has been one-hot coded. In this case, the number was seven:

```
y_test[0]
array([ 0.,   0.,   0.,   0.,   0.,   0.,   0.,   1.,   0.,   0.])
```

15. Now, we need to design our simple feedforward neural network with an input layer using the `sigmoid` activation function and 64 neurons. We will add a `softmax` function to the output layer, which does the classification by giving probabilities of the classified label:

```
model=Sequential()
model.add(Dense(64,activation='sigmoid', input_shape=(784,)))
model.add(Dense(10,activation='softmax'))
```

16. We can look at the structure of the neural network we just designed using the `summary()` function, which is a simple network with an input layer of 64 neurons and an output layer with 10 neurons. The output layer has 10 neurons we have 10 class labels to predict/classify (zero to nine):

```
model.summary()
```

Layer (type)	Output Shape	Param #
dense_1 (Dense)	(None, 64)	50240
dense_2 (Dense)	(None, 10)	650

```
Total params: 50,890
Trainable params: 50,890
Non-trainable params: 0
```

17. Next, we need to configure the model to use an optimizer, a cost function, and a metric to determine accuracy. Here, the optimizer that's being used is **Scalar Gradient Descent (SGD)** with a learning rate of 0.01. The loss function that's being used is the algebraic **Mean Squared Error (MSE)**, and the metric to measure the correctness of the model is `accuracy`, which is the probability score:

```
model.compile(loss='mean_squared_error',
optimizer=SGD(lr=0.01),metrics['accuracy'])
```

18. Now, we are ready to train the model. We want it to use 128 samples for every iteration of learning through the network, indicated by `batch_size`. We want each sample to iterate at least 200 times throughout the network, which is indicated by `epochs`. Also, we indicate the training and validation sets to be used. `Verbose` controls the output prints on the console:

```
model.fit(X_train,y_train,batch_size=128,epochs=200,
verbose=1,validation_data =(X_test,y_test))
```

19. Train on 60,000 samples, and then validate on 10,000 samples:

```
Epoch 1/200
60000/60000 [==============================] - 1s - loss: 0.0915 -
acc: 0.0895 - val_loss: 0.0911 - val_acc: 0.0955
Epoch 2/200
    .
    .
    .
60000/60000 [==============================] - 1s - loss: 0.0908 -
acc:
0.8579 - val_loss: 0.0274 - val_acc: 0.8649
Epoch 199/200
60000/60000 [==============================] - 1s - loss: 0.0283 -
acc: 0.8585 - val_loss: 0.0273 - val_acc: 0.8656
Epoch 200/200
60000/60000 [==============================] - 1s - loss: 0.0282 -
acc: 0.8587 - val_loss: 0.0272 - val_acc: 0.8658
<keras.callbacks.History at 0x7f308e68be48>
```

20. Finally, we can evaluate the model and how well the model predicts on the test dataset:

```
model.evaluate(X_test,y_test)
9472/10000 [=============================>..] - ETA: 0s
[0.027176343995332718, 0.86580000000000001]
```

This can be interpreted as having an error rate (MSE) of 0.027 and an accuracy of 0.865, which means it predicted the right label 86% of the time on the test dataset.

Building a feedforward neural network to recognize handwritten digits, version two

In the previous section, we built a very simple neural network with just an input and output layer. This simple neural network gave us an accuracy of 86%. Let's see if we can improve this accuracy further by building a neural network that is a little deeper than the previous version:

1. Let's do this on a new notebook. Loading the dataset and data pre-processing will be the same as in the previous section:

```
import numpy as np
np.random.seed(42)
import keras
from keras.datasets import mnist
from keras.models import Sequential
from keras.layers import Dense
from keras.optimizers import SG
#loading and pre-processing data
(X_train,y_train), (X_test,y_test)= mnist.load_data()
X_train= X_train.reshape( 60000, 784). astype('float32')
X_test =X_test.reshape(10000,784).astype('float32')
X_train/=255
X_test/=255
```

2. The design of the neural network is slightly different from the previous version. We will add a hidden layer with 64 neurons to the network, along with the input and output layers:

```
model=Sequential()
model.add(Dense(64,activation='relu', input_shape=(784,)))
model.add(Dense(64,activation='relu'))
model.add(Dense(10,activation='softmax'))
```

3. Also, we will use the `relu` activation function for the input and hidden layer instead of the `sigmoid` function we used previously.

4. We can inspect the model design and architecture as follows:

```
model.summary()
```

Layer (type)	Output Shape	Param #
dense_1 (Dense)	(None, 64)	50240
dense_2 (Dense)	(None, 64)	4160
dense_3 (Dense)	(None, 10)	650

```
Total params: 55,050
Trainable params: 55,050
Non-trainable params: 0
```

5. Next, we will configure the model to use the derivative `categorical_crossentropy` cost function rather than MSE. Also, the learning rate is increased from 0.01 to 0.1:

```
model.compile(loss='categorical_crossentropy',optimizer=SGD(lr=0.1)
,
metrics =['accuracy'])
```

6. Now, we will train the model, like we did in the previous examples:

```
model.fit(X_train,y_train,batch_size=128,epochs=200,verbose=1,valid
ation_data =(X_test,y_test))
```

7. Train on 60,000 samples and validate on 10,000 samples:

```
Epoch 1/200
60000/60000 [==============================] - 1s - loss: 0.4785 -
acc: 0.8642 - val_loss: 0.2507 - val_acc: 0.9255
Epoch 2/200
60000/60000 [==============================] - 1s - loss: 0.2245 -
acc: 0.9354 - val_loss: 0.1930 - val_acc: 0.9436
 .
 .
 .
60000/60000 [==============================] - 1s - loss:
4.8932e-04 - acc: 1.0000 - val_loss: 0.1241 - val_acc: 0.9774
<keras.callbacks.History at 0x7f3096adadd8>
```

As you can see, there is an increase in accuracy compared to the model we built in the first version.

Building a deeper neural network

In this section, we will use the concepts we learned about in this chapter to build a deeper neural network to classify handwritten digits:

1. We will start with a new notebook and then load the required dependencies:

```python
import numpy as np
np.random.seed(42)
import keras
from keras.datasets import mnist
from keras.models import Sequential
from keras.layers import Dense
from keras.layers import Dropout
# new!
from keras.layers.normalization
# new!
import BatchNormalization
# new!
from keras import regularizers
# new!
from keras.optimizers import SGD
```

2. We will now load and pre-process the data:

```python
(X_train,y_train),(X_test,y_test)= mnist.load_data()
X_train= X_train.reshape(60000,784).astype('float32')
X_test= X_test.reshape(10000,784).astype('float32')
X_train/=255
X_test/=255
n_classes=10
y_train=keras.utils.to_categorical(y_train,n_classes)
y_test =keras.utils.to_categorical(y_test,n_classes)
```

3. Now, we will design a deeper neural architecture with measures to take care of overfitting and to provide better generalization:

```python
model=Sequential()
model.add(Dense(64,activation='relu',input_shape=(784,)))
model.add(BatchNormalization())
model.add(Dropout(0.5))
model.add(Dense(64,activation='relu'))
model.add(BatchNormalization())
```

```
model.add(Dropout(0.5))
model.add(Dense(10,activation='softmax'))
model.summary()
```

Layer (type)	Output Shape	Param #
dense_1 (Dense)	(None, 64)	50240
batch_normalization_1 (Batch	(None, 64)	256
dropout_1 (Dropout)	(None, 64)	0
dense_2 (Dense)	(None, 64)	4160
batch_normalization_2 (Batch	(None, 64)	256
dropout_2 (Dropout)	(None, 64)	0
dense_3 (Dense)	(None, 10)	650

```
Total params: 55,562
Trainable params: 55,306
Non-trainable params:
256
```

4. This time, we will configure the model using an adam optimizer:

```
model.compile(loss='categorical_crossentropy',optimizer='adam',metr
ics=['accuracy'])
```

5. Now, we will post that we will train the model for 200 epochs at a batch size of 128:

```
model.fit(X_train, y_train, batch_size= 128, epochs= 200, verbose=
1, validation_data= (X_test,y_test))
```

6. Train on 60,000 samples and validate on 10,000 samples:

```
Epoch 1/200
60000/60000 [==============================] - 3s - loss: 0.8586 -
acc: 0.7308 - val_loss: 0.2594 - val_acc: 0.9230
Epoch 2/200
60000/60000 [==============================] - 2s - loss: 0.4370 -
acc: 0.8721 - val_loss: 0.2086 - val_acc: 0.9363
  .
  .
  .
Epoch 200/200
```

```
60000/60000 [==============================] - 2s - loss: 0.1323 -
acc: 0.9589 - val_loss: 0.1136 - val_acc: 0.9690
<keras.callbacks.History at 0x7f321175a748>
```

Introduction to Computer Vision

Computer Vision can be defined as the subset of AI where we can teach a computer to *see*. We cannot just add a camera to a machine in order for it to *see*. For a machine to actually view the world like people or animals do, it relies on Computer Vision and image recognition techniques. Reading barcodes and face recognition are examples of Computer Vision. Computer Vision can be described as that part of the human brain that processes the information that's perceived by the eyes, nothing else.

Image recognition is one of the interesting uses of Computer Vision from an AI standpoint. The input that is received through Computer Vision on the machine is interpreted by the image recognition system, and based on what it sees, the output is classified.

In other words, we use our eyes to capture the objects around us, and those objects/images are processed in our brain, which allows us to visualize the world around us. This capability is given by Computer Vision to machines. Computer Vision is responsible for automatically extracting, analyzing, and understanding the required information from the videos or images that are fed in as input.

There are various Computer Vision application, and they are used in the following scenerios:

- Augmented reality
- Robotics
- Biometrics
- Pollution monitoring
- Agriculture
- Medical image analysis
- Forensics
- Geoscience
- Autonomous vehicles
- Image restoration
- Process control
- Character recognition
- Remote sensing

- Gesture analysis
- Security and surveillance
- Face recognition
- Transport
- Retail
- Industrial quality inspection

Machine learning for Computer Vision

It's important to use the appropriate ML theories and tools, which will be very helpful when we need to develop various applications that involve classifying images, detecting objects, and so on. Utilizing these theories to create computer vision applications requires an understanding of some basic machine learning concepts.

Conferences help on Computer Vision

Some of the conferences to look for latest research and applications are as follows:

- **Conference on Computer Vision and Pattern Recognition (CVPR)** is held every year and is one of the popular conferences with research papers ranging from both theory and application across a wide domain
- **International Conference on Computer Vision (ICCV)**is another major conference held every other year attracting one of the best research papers
- **Special Interest Group on Computer Graphics (SIGGRAPH)** and interactive techniques though more on computer graphics domain has several applications papers that utilizes computer vision techniques.

Other notable conferences include **Neural Information Processing Systems (NIPS), International Conference on Machine Learning (ICML), Asian Conference on Computer Vision (ACCV), European Conference on Computer Vision (ECCV)**, and so on.

Summary

In this chapter, we built a feedforward neural network to recognize handwritten digits in two versions. Then, we built a neural network to classify handwritten digits, and, finally we gave a short introduction to Computer Vision.

In the next chapter, we will build a Machine Vision mobile app to classify flower species and retrieve the necessary information.

Further reading

For in-depth knowledge on computer vision, do refer the following Packt books:

- *Deep Learning for Computer Vision* by Rajalingappaa Shanmugamani
- *Practical Computer Vision* by Abhinav Dadhich

4
Building a Machine Vision Mobile App to Classify Flower Species

In this chapter, we are going to use the theoretical knowledge we have learned in previous chapters to create a mobile application that will classify a specific species of flower. By utilizing use your mobile camera and pointing it at a flower, the application will analyze the image and make its best educated guess as to the species of that flower. This is where we put to work the understanding we have gained about the workings of a **convolutional neural network** (**CNN**). We will also learn a bit more about using TensorFlow as well as some tools such as TensorBoard. But before we dive in too deep, let's talk about a few things first.

Throughout this chapter we use terms that may not be familiar to all, so let's make sure we're all on the same page as to what they mean.

In this chapter, we will cover the following topics:

- CoreML versus TensorFlow Lite
- What is MobileNet
- Datasets for image classification
- Creating your own image dataset
- Using TensorFlow to build the model
- Running TensorBoard

CoreML versus TensorFlow Lite

In the machine learning world, there are two efforts (as of the time of this writing) taking place in order to improve the mobile AI experience. Instead of offloading AI or ML processing to the cloud and a data center, the faster option would be to process data on the device itself. In order to do this, the model must already be pre-trained, which means that there is a chance that it is not trained for exactly what you are going to use it for.

In this space, Apple's effort (iOS) is called **Core ML**, and Google's (Android) is called **TensorFlow Lite**. Let's talk briefly about both.

CoreML

The CoreML framework from Apple provides a large selection of neural network types. This allows developers to be able to experiment with different designs when developing their apps. Camera and microphone data are just two area which can be leveraged for things like image recognition, natural language processing, and more. There are several pre-trained models that developers can use straight out of the box, and tweak as necessary for their application.

TensorFlow Lite

TensorFlow Lite is what is known as a local-device version of TensorFlow, meaning it is designed to run on your mobile device itself. As of the time of this writing it is still in pre-release status, so an exact comparison to CoreML is difficult. We will have to wait and see what the final offering provides. For now, simply be aware there are two options for mobile device-local AI and machine learning.

What is MobileNet?

Before we dive in too deep, let us first talk about a term you will hear used quite a bit in this chapter, **MobileNets**. What is a MobileNet you might ask? Simply put, it's an architecture which is designed specifically for mobile and embedded vision-based applications. On such devices there is a lack of computing power available for such processing, which therefore increases the need for a better solution that one used on a desktop environment.

The **MobileNet** architecture was proposed by Google, and briefly:

1. Uses depth-wise separable convolutions. This significantly reduces the number of parameters when compared to a neural network using normal convolutions with the same depth. The result is what is known as a **light-weight deep neural network**.

2. **Depth-wise convolution**, followed by **Pointwise convolution**, replaces the normal convolution process.

In order to simplify things, we are going to break down this chapter into the following two sections:

- **Datasets for Image Classification**: In this section we will explore the various datasets (all of which are available online) that can be used for image classification. We will also address the issue of how to create our own datasets, if necessary.

- **Using TensorFlow to Build the Model**: In this section we will use TensorFlow to train our classification model. We do this by using a pretrained model called **MobileNet**. MobileNets are a family of mobile-first computer vision models for TensorFlow, designed to maximize accuracy while considering the restricted resources available for an on-device or embedded application.

- In addition, we will look at converting the output model into a `.tflite` format, which can be used within other mobile or embedded devices. TFLite stands for TensorFlow Lite. You can learn more about TensorFlow Lite via any internet search engine.

Datasets for image classification

For our flower classification example, we will be using the University of Oxford's **Visual Geometry Group (VGG)** image dataset collection. The collection can be accessed at `http://www.robots.ox.ac.uk/~vgg/data/`.

The VGG is the same department that won previous ImageNet competitions. The pretrained models, such as VGG14 and VGG16, were built by this department and they won in 2014 and 2016, respectively. These datasets are used by the VGG to train and evaluate the models that they build.

The flower dataset can be found in the **Fine-Grain Recognition Datasets** section of the page, along with textures and pet datasets. Click on **Flower Category Datasets**, or use the following link to access the flower datasets from VGG, http://www.robots.ox.ac.uk/~vgg/data/flowers/.

Here, you can find two datasets, one with 17 different species of flowers, and the other with 102 different species of flowers. You can choose either one based on their ease of use for the tutorial, or based on the kind of processing that is available at your disposal.

 Using a larger dataset means that the training will take longer, and so will the data processing before training; therefore, we recommend that you choose wisely.

Here is a subset of the images you will find here. As you will see, the folder names match up identically with those we will use a bit later on in this chapter:

Aside from the images we talked about above, here are several additional links that you can use to get image data for similar classification use cases should you ever desire to use them:

- **CVonline datasets**: `http://homepages.inf.ed.ac.uk/rbf/CVonline/Imagedbase.htm`
- **CVpapers datasets**: `http://www.cvpapers.com/datasets.html`
- **Image datasets**: `http://wiki.fast.ai/index.php/Image_Datasets`
- **Deep learning datasets**: `http://deeplearning.net/datasets/`
- **COCO datsets**: `http://cocodataset.org/#home`
- **ImageNet datasets**: `http://www.image-net.org/`
- **Open Images datasets**: `https://storage.googleapis.com/openimages/web/index.html`
- **Kaggle datasets**: `https://www.kaggle.com/datasets?sortBy=relevancegroup=featuredsearch=image`
- **Open datasets**: `https://skymind.ai/wiki/open-datasets`
- **Wikipedia**: `https://en.wikipedia.org/wiki/List_of_datasets_for_machine_learning_research#Object_detection_and_recognition`

Creating your own image dataset using Google images

Let's say, for whatever reason, we need to determine what kind of dog a picture is of, but we do not have any pictures readily available on our computer. What can we do? Well, perhaps the easiest approach is to open Google Chrome and search for the images online.

As an example, let's say we are interested in Doberman dogs. Just open Google Chrome and search for **doberman** pictures as shown below:

1. **Perform a search for Doberman pictures**: On searching, following the result were obtained:

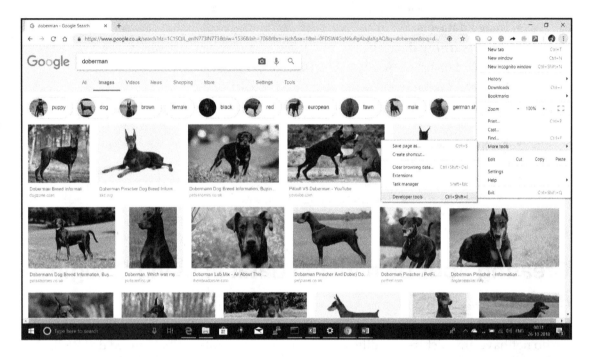

2. **Open the JavaScript console**: You can find the JavaScript Console in Chrome in the top-right menu:

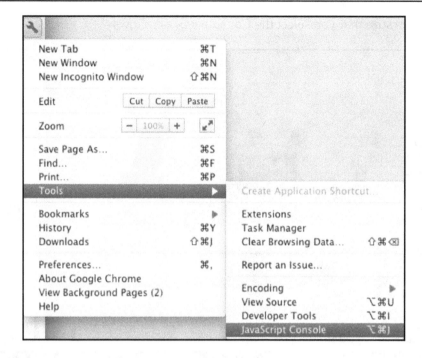

Click on **More tools** and then **Developer tools**:

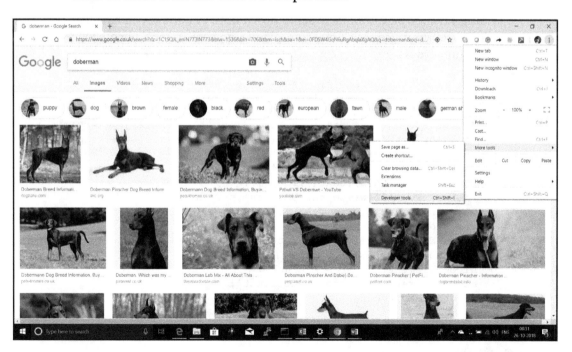

Make sure that you select the **Console** tab, as follows:

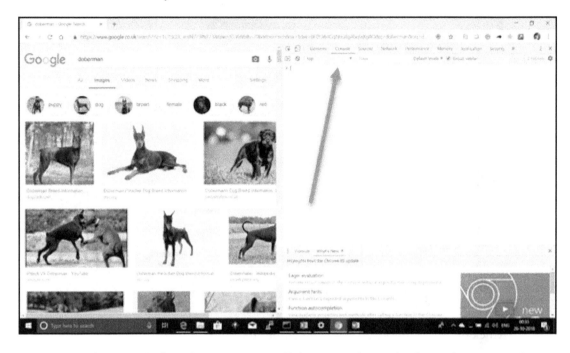

3. **Using JavaScript**: Continue to scroll down until you think you have enough images for your use case. Once this is done, go back to the **Console** tab in **Developer tools**, and then copy and paste the following script:

```
//the jquery  is pulled down in the JavaScript console
var script = document.createElement('script');
script.src =
"https://ajax.googleapis.com/ajax/libs/jquery/2.2.0/jquery.min.js";
document.getElementsByTagName('head')[0].appendChild(script);
//Let us get the URLs
var urls = $('.rg_di .rg_meta').map(function() { return
JSON.parse($(this).text()).ou; });
// Now, we will write the URLs one per line to file
var textToSave = urls.toArray().join('\n');
var hiddenElement = document.createElement('a');
hiddenElement.href = 'data:attachment/text,' +
encodeURI(textToSave);
hiddenElement.target = '_blank';
hiddenElement.download = 'urls.txt';
hiddenElement.click();
```

This code snippet collects all the image URLs and saves them to a file called urls.txt in your default Downloads directory.

4. **Use Python to download the images**: Now, we will use Python to read the URLs of the images from urls.txt and download all the images into a folder:

This can be done easily by following the following steps:

1. Open a Python notebook and copy and paste the following code to download the images:

```
# We will start by importing the required pacages
from imutils import paths
import argparse
import requests
import cv2
import os
```

2. After importing, start constructing the arguments, and after constructing parsing the arguments is important:

```
ap = argparse.ArgumentParser()
ap.add_argument("-u", "--urls", required=True,
help="path to file containing image URLs")
ap.add_argument("-o", "--output", required=True,
help="path to output directory of images")
args = vars(ap.parse_args())
```

3. The next step includes grabbing the list of URLs from the input file counting total number of images downloaded:

```
rows = open(args["urls"]).read().strip().split("\n")
total = 0
# URLs are looped in
for url in rows:
try:
# Try downloading the image
r = requests.get(url, timeout=60)
#The image is then saved to the disk
p = os.path.sep.join([args["output"], "{}.jpg".format(
```

```
str(total).zfill(8))])
f = open(p, "wb")
f.write(r.content)
f.close()
#The counter is updated
print("[INFO] downloaded: {}".format(p))
total += 1
```

4. During the download process, the exceptions that are thrown need to be handled:

```
print("[INFO] error downloading {}...skipping".format(p))
```

5. The image paths that are downloaded need to be looped over:

```
for imagePath in paths.list_images(args["output"])
```

6. Now, decide whether the image should be deleted or not and accordingly initialize:

```
delete = False
```

7. The image needs to be loaded. Let's try to do that:

```
image = cv2.imread(imagePath)
```

8. If we weren't able to load the image properly, since the image is None, then it should be deleted from the disk:

```
if image is None:
delete = True
```

9. Also, if OpenCV was unable to load the image, it means the image is corrupt and should be deleted:

```
except:
print("Except")
delete = True
```

10. Give a final check and see whether the image was deleted:

```
if delete:
print("[INFO] deleting {}".format(imagePath))
os.remove(imagePath)
```

11. With that complete, let's download this notebook as a Python file and name it `image_download.py`. Make sure that you place the `urls.txt` file in the same folder as the Python file that you just created. This is very important.

12. Next, we need to execute the Python file we just created. We will do so by using the command line as shown here (make sure your `path` variable points to your Python location):

```
Image_download.py --urls urls.txt --output Doberman
```

By executing this command, the images will be downloaded to the folder named Doberman. Once this has been completed, you should see all the images of the Doberman that you viewed in Google Chrome, like what is shown in the following image:

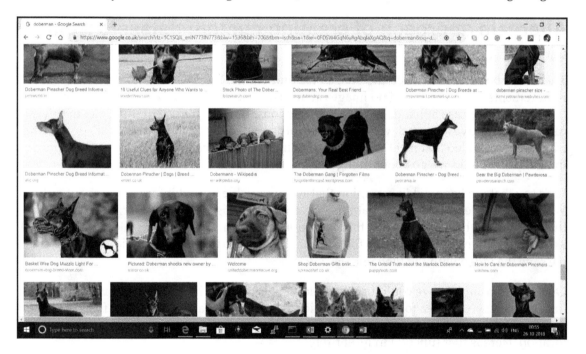

Select the required folder for saving the images as shown:

That's it we now have a folder full of Doberman images. The same method can be applied to create a folder of any other type of category that we may need.

There may be a number of images that are part of the Google image results that are not desirable. Ensure that you browse through the images and remove any unwanted images.

Alternate approach of creating custom datasets from videos

There may be occasions when the images we find via the Internet do not satisfy our requirements or, we may find no images at all. This could be because of the uniqueness of the data, the use case at hand, copyright restrictions, the required resolution, etc. In this case, an alternative approach would be to record a video of the object you need, extract the frames of that video that meet your requirements, and save each frame as an individual image. How would we go about doing that?

Let's say that we have a skin condition that we are unable to find information about online. We need to somehow classify what this skin condition might be. However, in order to do this, we need to have an image of this condition. Accordingly, we could take a video of that skin condition and save the video file to a file. For the purposes of discussion, let's say that we save the video with the filename `myvideo.mp4`.

Once this is complete, we could use the following Python script to break the video into images and save it into a folder. This function will take the path of the video file, break it into frames based on frequency, and save the corresponding images to a specified output location. Here is that function in its entirety:

```
import sys
import argparse
import os
import cv2
import numpy as np
print(cv2.__version__)
```

This function takes the path of the video file, breaks it into frames based on frequency, and saves the corresponding images to a specified output location:

```
def extractImages(pathIn, pathOut):
count = 0
vidcap = cv2.VideoCapture(pathIn)
success,image = vidcap.read()
success = True
while success:
vidcap.set(cv2.CAP_PROP_POS_MSEC,(count*10)) # Adjust frequency of frames
here
success,image = vidcap.read()
print ('Read a new frame: ', success)
#Once we identify the last frame, stop there
image_last = cv2.imread("frame{}.png".format(count-1))
if np.array_equal(image,image_last):
break
cv2.imwrite( os.path.join("frames","frame{:d}.jpg".format(count)), image) #
save frame as JPEG file
count = count + 1
pathIn = "myvideo.mp4"
pathOut = ""
extractImages(pathIn, pathOut)
```

As mentioned above, this will save every frame of the video in the current folder based on the frequency set. After running this script, you now will have created your image dataset and be able to use the images you need.

Building your model using TensorFlow

Now that we have seen several methods of obtaining the images we need, or, in the absence of any, creating our own, we will now use TensorFlow to create the classification model for our flower use case:

1. **Creating the folder structure**: To start with, let's create the folder structure that's required for our flower classification use case. First, create a main folder called `image_classification`. Within the `image_classification` folder, create two folders: `images` and `tf_files`. The `images` folder will contain the images that are required for model training, and the `tf_files` folder will hold all the generated TensorFlow-specific files during runtime.

2. **Downloading the images**: Next, we need to download the images that are specific to our use case. Using the example of **Flowers**, our images will come from the VGG datasets page we discussed earlier.

Please feel free to use your own datasets, but make sure that each category is in its own separate folder. Place the downloaded image dataset within the images folder.

For example, the complete folder structure will look like this:

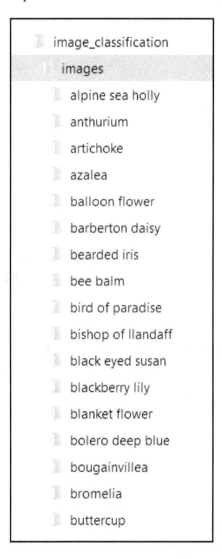

3. **Creating the Python script**: In this step, we will create the TensorFlow code that is required to build our model. Create a Python file named `retrain.py` within the main `image_classification` folder.

Once this is complete, the following code block should be copied and used. Below we have broken out the process into several steps in order to describe what is taking place:

1. The following code block is the complete script that goes into `retrain.py`:

```
from __future__ import absolute_import
from __future__ import division
from __future__ import print_function
import argparse
import collections
from datetime import datetime
import hashlib
import os.path
import random
import re
import sys
import tarfile
import numpy as np
from six.moves import urllib
import tensorflow as tf
from tensorflow.python.framework import graph_util
from tensorflow.python.framework import tensor_shape
from tensorflow.python.platform import gfile
from tensorflow.python.util import compat
FLAGS = None
MAX_NUM_IMAGES_PER_CLASS = 2 ** 27 - 1 # ~134M
```

2. Next, we need to prepare the images so that they can be trained, validated, and tested:

```
result = collections.OrderedDict()
sub_dirs = [
os.path.join(image_dir,item)
for item in gfile.ListDirectory(image_dir)]
sub_dirs = sorted(item for item in sub_dirs
if gfile.IsDirectory(item))
for sub_dir in sub_dirs:
```

The first thing we are going to do is to retrieve the images from the directory path where they are stored. We will use the images to create the model graph using the model that you previously downloaded and installed.

The next step is to bottleneck the array initialization by creating what is known as **bottleneck files**. **Bottleneck** is an informal term used for the layer just before the final output layer that does the actual classification. (TensorFlow Hub calls this an **image feature vector**.) This layer has been trained to output a set of values that's good enough for the classifier to use in order to distinguish between all the classes it's been asked to recognize. This means that it must be a meaningful and compact summary of the images, since it must contain enough information for the classifier to make a good choice in a very small set of values.

It's important that we have bottleneck values for each image. If the bottleneck values aren't available for each image, we will have to create them manually because these values will be required in the future when training the images. It is highly recommended to cache these values in order to speed up processing time later. Because every image is reused multiple times during training, and calculating each bottleneck takes a significant amount of time, it speeds things up to cache these bottleneck values on disk to avoid repeated recalculated. By default, bottlenecks are stored in the `/tmp/bottleneck` directory (unless a new directory was specified as an argument).

When we retrieve the bottleneck values, we will do so based upon the filenames of images that are stored in the cache. If distortions were applied to images, there might be difficulty in retrieving the bottleneck values. The biggest disadvantage of enabling distortions in our script is that the bottleneck caching is no longer useful, since input images are never reused exactly. This directly correlates to a longer training process time, so it is highly recommended this happens once you have a model that you are reasonably happy with. Should you experience problems, we have supplied a method of getting the values for images which have distortions supplied as a part of the GitHub repository for this book.

 Please note that we materialize the distorted image data as a NumPy array first.

Next, we need to send the running inference on the image. This requires a trained object detection model and is done by using two memory copies.

Our next step will be to apply distortion to the images. Distortions such as cropping, scaling and brightness are supplied as percentage values which control how much of each distortion is applied to each image. It's reasonable to start with values of 5 or 10 for each of them and then experiment from there to see which/what helps and what does not.

We next need to summarize our model based on accuracy and loss. We will use TensorBoard visualizations to analyze it. If you do not already know, TensorFlow offers a suite of visualization tools called TensorBoard which allows you to visualize your TensorFlow graph, plot variables about the execution, and show additional data like images that pass through it. The following is an example TensorBoard dashboard:

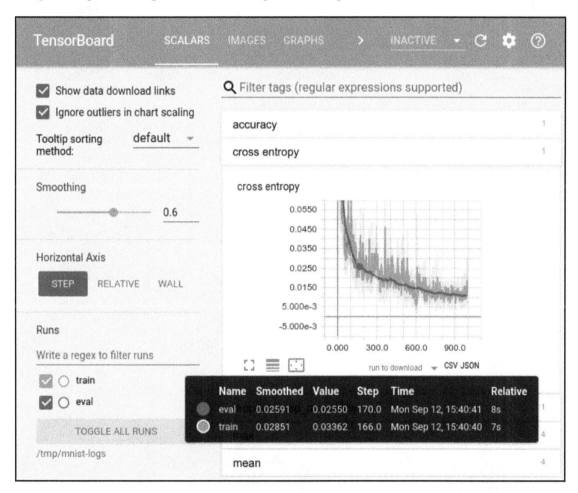

Our next step will be to save the model to a file, as well as setting up a directory path to write summaries for the TensorBoard.

At this point we should point out the `create_model_info` function, that will return the model information. In our example below, we handle both MobileNet and Inception_v3 architectures. You will see later how we handle any other architecture but these:

```
def create_model_info(architecture):
architecture = architecture.lower()
if architecture == 'inception_v3':
# pylint: disable=line-too-long
data_url =
'http://download.tensorflow.org/models/image/imagenet/inception-2015-12-05.
tgz'
# pylint: enable=line-too-long
bottleneck_tensor_name = 'pool_3/_reshape:0'
bottleneck_tensor_size = 2048
input_width = 299
input_height = 299
input_depth = 3
resized_input_tensor_name = 'Mul:0'
model_file_name = 'classify_image_graph_def.pb'
input_mean = 128
input_std = 128
elif architecture.startswith('mobilenet_'):
parts = architecture.split('_')
if len(parts) != 3 and len(parts) != 4:
tf.logging.error("Couldn't understand architecture name '%s'",
architecture)
return None
version_string = parts[1]
if (version_string != '1.0' and version_string != '0.75' and
version_string != '0.50' and version_string != '0.25'):
tf.logging.error(
"""The Mobilenet version should be '1.0', '0.75', '0.50', or '0.25',
but found '%s' for architecture '%s'""",
version_string, architecture)
return None
size_string = parts[2]
if (size_string != '224' and size_string != '192' and
size_string != '160' and size_string != '128'):
tf.logging.error(
"""The Mobilenet input size should be '224', '192', '160', or '128',
but found '%s' for architecture '%s'""",
size_string, architecture)
return None
if len(parts) == 3:
is_quantized = False
```

If the above argument turns out to be false, this means that we encountered an architecture which we were not expecting. If this happens, we will need to execute the following code block to obtain the result. In this instance we are not dealing with either MobileNet or Inception_V3 and will default to using version 1 of MobileNet:

```
else:
if parts[3] != 'quantized':
tf.logging.error(
"Couldn't understand architecture suffix '%s' for '%s'", parts[3],
architecture)
return None
is_quantized = True
data_url = 'http://download.tensorflow.org/models/mobilenet_v1_'
data_url += version_string + '_' + size_string + '_frozen.tgz'
bottleneck_tensor_name = 'MobilenetV1/Predictions/Reshape:0'
bottleneck_tensor_size = 1001
input_width = int(size_string)
input_height = int(size_string)
input_depth = 3
resized_input_tensor_name = 'input:0'
if is_quantized:
model_base_name = 'quantized_graph.pb'
else:
model_base_name = 'frozen_graph.pb'
model_dir_name = 'mobilenet_v1_' + version_string + '_' + size_string
model_file_name = os.path.join(model_dir_name, model_base_name)
input_mean = 127.5
input_std = 127.5
else:
tf.logging.error("Couldn't understand architecture name '%s'",
architecture)
raise ValueError('Unknown architecture', architecture)
return {
'data_url': data_url,
'bottleneck_tensor_name': bottleneck_tensor_name,
'bottleneck_tensor_size': bottleneck_tensor_size,
'input_width': input_width,
'input_height': input_height,
'input_depth': input_depth,
'resized_input_tensor_name': resized_input_tensor_name,
'model_file_name': model_file_name,
'input_mean': input_mean,
'input_std': input_std,
}
================================================================
```

Another important point we should note is that we will need to decode the image JPEG data after processing. The following function, `add_jpeg_decoding`, is a complete code snippet which does this by calling the `tf.image.decode_jpeg` function:

```
def add_jpeg_decoding(input_width, input_height, input_depth, input_mean,
input_std):
jpeg_data = tf.placeholder(tf.string, name='DecodeJPGInput')
decoded_image = tf.image.decode_jpeg(jpeg_data, channels=input_depth)
decoded_image_as_float = tf.cast(decoded_image, dtype=tf.float32)
decoded_image_4d = tf.expand_dims(decoded_image_as_float, 0)
resize_shape = tf.stack([input_height, input_width])
resize_shape_as_int = tf.cast(resize_shape, dtype=tf.int32)
resized_image = tf.image.resize_bilinear(decoded_image_4d,
resize_shape_as_int)
offset_image = tf.subtract(resized_image, input_mean)
mul_image = tf.multiply(offset_image, 1.0 / input_std)
return jpeg_data, mul_image
```

And here, in all its glory is our `main` function. Basically we do the following:

- Set our logging level to `INFO`
- Prepare the file system for usage
- Create our model information
- Download and extract our data

```
def main(_):
tf.logging.set_verbosity(tf.logging.INFO)
prepare_file_system()
model_info = create_model_info(FLAGS.architecture)
if not model_info:
tf.logging.error('Did not recognize architecture flag')
return -1
maybe_download_and_extract(model_info['data_url'])
graph, bottleneck_tensor, resized_image_tensor = (
create_model_graph(model_info))
image_lists = create_image_lists(FLAGS.image_dir, FLAGS.testing_percentage,
FLAGS.validation_percentage)
```

The preceding `retrain.py` file is available for download as part of the assets within this book.

Running TensorBoard

To run TensorBoard, use the following command:

```
tensorboard --logdir=path/to/log-directory
```

Where `logdir` points to the directory where serialized data is contained. If this directory contains subdirectories which also contain serialized data, TensorBoard will visualize the data from all of those runs. Once TensorBoard is running, navigate your web browser to `localhost:6006` to view the TensorBoard and its associated data.

For those wanting or needing to learn more about TensorBoard, please check out the following tutorial at `https://www.tensorflow.org/tensorboard/r1/summaries`.

Summary

In this chapter we have accomplished a lot in this small chapter. We began the chapter with understanding the various datasets that are available for image classification, as well as how we could obtain or create images if we could not find any that met our requirements. Next, then divided the chapter into two distinct sections. In the first section we learned about creating our own image dataset. In the second section we learned how to use TensorFlow to build the model.

In the next chapter, we are going to extend our TensorFlow knowledge even further by using various TensorFlow libraries to build a machine learning model which will predict body damage done to a car.

5
Building an ML Model to Predict Car Damage Using TensorFlow

In this chapter, we will build a system that detects the level of damage that's been done to a vehicle by analyzing photographs using **transfer learning**. A solution like this will be helpful in reducing the cost of insurance claims, as well as making the process simpler for vehicle owners. If the system is implemented properly, in an ideal scenario, the user will upload a bunch of photographs of the damaged vehicle, the photos will go through damage assessment, and the insurance claim will be processed automatically.

There are a lot of risks and challenges involved in implementing a perfect solution for this use case. To start with, there are multiple unknown conditions that could have caused damage to the car. We are not aware of the outdoor environment, surrounding objects, light in the area, and the quality of the vehicle before the incident. Passing through all these hurdles and figuring out a common solution for the problem is challenging. This is a common problem across any computer vision-based scenario.

In this chapter, we will cover the following topics:

- Transfer learning basics
- Image dataset collections
- Setting up a web application
- Training our own TensorFlow model
- Building a web app that consumes the model

Transfer learning basics

To implement the car damage prediction system, we are going to build our own TensorFlow-based **machine learning** (**ML**) model for the vehicle datasets. Millions of parameters are needed with modern recognition models. We need a lot of time and data to train a new model from scratch, as well as hundreds of **Graphical Processing Units** (**GPUs**) or **Tensor Processing Units** (**TPUs**) that run for hours.

Transfer learning makes this task easier by taking an existing model that is already trained and reusing it on a new model. In our example, we will use the feature extraction capabilities from the **MobileNet** model and train our own classifiers on top of it. Even if we don't get 100% accuracy, this works best in a lot of cases, especially on a mobile phone where we don't have heavy resources. We can easily train this model on a typical laptop for a few hours, even without a GPU. The model was built on a MacBook Pro with a 2.6 GHz Intel i5 processor and 8 GB RAM.

Transfer learning is one of the most popular approaches in deep learning, where a model that's been developed for one task is reused for another model on a different task. Here, pre-trained models are used as a first step in computer vision-based tasks or **natural language processing** (**NLP**) based tasks, provided we have very limited computational resources and time.

In a typical computer vision-based problem, neural networks try to detect edges in their initial level layers, shapes in the middle level layers, and more specific features in the final level layers. With transfer learning, we will use the initial and middle level layers and only retrain the final level layers.

For example, if we have a model that's trained to recognize an apple from the input image, it could be reused to detect water bottles. In the initial layers, the model has been trained to recognize objects, so we will retrain only the final level layers. In that way, our model will learn what will differentiate water bottles from other objects. This process can be seen in the following diagram:

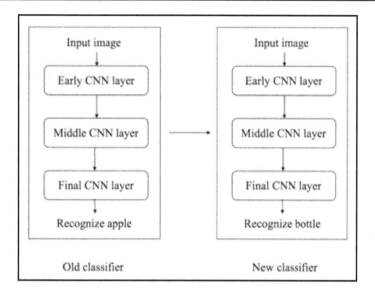

Typically, we need a lot of data to train our model, but most of the time we will not have enough relevant data. That is where transfer learning comes into the picture, and is where you can train your model with very little data.

If your old classifier was developed and trained using TensorFlow, you can reuse the same model to retrain a few of the layers for your new classifier. This will work perfectly, but only if the features that were learned from the old task are more generic in nature. For example, you can't reuse a model that was developed for a text classifier on an image classification-based task. Also, the input data size should match for both the models. If the size doesn't match, we need to add an additional preprocessing step to resize the input data accordingly.

Approaches to transfer learning

Let's look into different approaches to transfer learning. There could be different names given to the approaches, but the concepts remain the same:

1. **Using a pre-trained model**: There are a lot of pre-trained models out there to satisfy your basic deep learning research. In this book, we have used a lot of pre-trained models from where we derive our results.

2. **Training a model for reuse**: Let's assume that you want to solve problem A, but you don't have enough data to achieve the results. To solve this issue, we have another problem, B, where we have enough data. In that case, we can develop a model for problem B, and use the model as a starting point for problem A. Whether we need to reuse all the layers or only a few layers is dependent on the type of problem that we are solving.

3. **Feature extraction**: With deep learning, we can extract the features of the dataset. Most of the time, the features are handcrafted by the developers.
Neural networks have the ability to learn which features you have to pass on, and which ones you don't. For example, we will only use the initial layers to detect the right representation of features, but we will not use the output because it might be more specific to one particular task. We will simply feed the data into our network and use one of the immediate middle level layers as the output layer.

With this, we will start building our model using TensorFlow.

Building the TensorFlow model

Building your own custom model requires following a step-by-step procedure. To begin, we are going to use the TensorFlow Hub to feed images using pre-trained models.

 To learn more about TensorFlow Hub, please refer to `https://www.tensorflow.org/hub`.

Installing TensorFlow

While writing this book, TensorFlow r1.13 was available. Revision 2.0.0 is also available on the alpha stage, but we will stay with a stable version. The TensorFlow Hub has a dependency on the TensorFlow library that can be installed with `pip`, as follows:

```
$ pip install tensorflow
$ pip install tensorflow-hub
```

When the `tensorflow` library is installed, we need to start collecting our image dataset before the training process starts. We need to look into a lot of things before we begin our training.

Training the images

In this section, we will collect the images and keep them organized under their respective folder categories.

A few common steps for choosing your own dataset of images are as follows:

1. First of all, you need at least 100 photos of each image category that you want to recognize. The accuracy of your model is directly proportional to the number of images in the set.

2. You need to make sure that you have more relevant images in the image set. For example, if you have taken an image set with a single color background let's say all the objects in the images have a white background and are shot indoors and users are trying to recognize objects with distracting backgrounds (for example, colorful backgrounds shot outdoors), this won't result in better accuracy.

3. Choose images with a variety of backgrounds. For example, if you are picking images with only two background colors, then your prediction will have a preference toward those two colors, rather than the object in the image.

4. Try to split bigger categories into smaller divisions. For example, instead of animal, you might use cat, dog, or tiger.

5. Make sure that you select all the input images that contain the objects that you are trying to recognize. For example, if you have a dog-recognizing app, we wouldn't use cars, buildings, or mountains as input images. In that case, it is better to have a separate classifier for the unrecognizable images.

6. Ensure that you label images properly. For example, labeling a flower as jasmine might have the whole plant in the picture or a person behind it. The accuracy of our algorithm will differ when there are distracting objects in the input images. Let's say you have taken a few food item images from Google Images. These images have reusable permissions, so always ensure that you have this when collecting images for your model. You can do this by searching any keyword from Google Images and filter the images based on labelled for reuse usage rights. You can find this option by clicking on tools beneath the search bar.

We have collected a few images from the internet for educational purposes in this chapter. This is discussed in further detail in the next section.

Building our own model

Here, we are going to build our own ML model using TensorFlow to analyze the damage level of the vehicle. We need to be careful in picking the dataset that will play a crucial part in the damage evaluation phase. Here are the steps that we are going to follow in order to build the model:

1. Find the image dataset of the damaged vehicles.
2. Categorize the images based on their damage levels. First, we need to identify that the object in the picture is actually a car. To do that, we need to have two categories of image sets that do and do not have cars in them. Then, we need to have three more categories to find the damage level of the cars categorized under high, medium, or low levels. Make sure that you have at least 1,000+ images under each of the five categories. Once the dataset is ready, we are ready to train our model.
3. We will train our model using TensorFlow.
4. We will build a web application to analyze the damage level of the vehicle.
5. Update the result.

Retraining with our own images

We will now use the `retrain.py` script in our project directory.

Download this script using `curl`, as follows:

```
mkdir -/Chapter5/images
cd -/Chapter5/images
curl -LO
https://github.com/tensorflow/hub/raw/master/examples/image_retraining/
retrain.py
python retrain.py --image_dir ./images/
```

There are a few parameters that have to be passed to the training script and looked into before the training starts.

Once our dataset is ready, we need to look into improving the results. We can do this by altering the number of steps in the learning process.

The simplest way to do this is by using the following code:

```
--how_many_training_steps = 4000
```

The rate of accuracy improvement slows down when the number of steps increases, and the accuracy will stop improving beyond a certain point. You can experiment with this and decide what works best for you.

Architecture

MobileNet is a smaller, low-power, low-latency model that's designed to meet the constraints of mobile devices. In our application, we have picked the following architecture from the MobileNet datasets as one of the parameters, as shown in the following code, for while we build the model, which has a better accuracy benchmark:

```
--architecture=" mobilenet_v2_1.4_224"
```

The power and latency of the network grows with the number of **Multiply-Accumulates (MACs)**, which measure the number of fused multiplication and addition operations, as follows:

Classification Checkpoint	MACs (M)	Parameters (M)	Top 1 Accuracy	Top 5 Accuracy	Mobile CPU (ms) Pixel 1
mobilenet_v2_1.4_224	582	6.06	75.0	92.5	138.0
mobilenet_v2_1.3_224	509	5.34	74.4	92.1	123.0
mobilenet_v2_1.0_224	300	3.47	71.8	91.0	73.8
mobilenet_v2_1.0_192	221	3.47	70.7	90.1	55.1
mobilenet_v2_1.0_160	154	3.47	68.8	89.0	40.2
mobilenet_v2_1.0_128	99	3.47	65.3	86.9	27.6
mobilenet_v2_1.0_96	56	3.47	60.3	83.2	17.6
mobilenet_v2_0.75_224	209	2.61	69.8	89.6	55.8
mobilenet_v2_0.75_192	153	2.61	68.7	88.9	41.6
mobilenet_v2_0.75_160	107	2.61	66.4	87.3	30.4
mobilenet_v2_0.75_128	69	2.61	63.2	85.3	21.9
mobilenet_v2_0.75_96	39	2.61	58.8	81.6	14.2
mobilenet_v2_0.5_224	97	1.95	65.4	86.4	28.7
mobilenet_v2_0.5_192	71	1.95	63.9	85.4	21.1
mobilenet_v2_0.5_160	50	1.95	61.0	83.2	14.9
mobilenet_v2_0.5_128	32	1.95	57.7	80.8	9.9
mobilenet_v2_0.5_96	18	1.95	51.2	75.8	6.4
mobilenet_v2_0.35_224	59	1.66	60.3	82.9	19.7
mobilenet_v2_0.35_192	43	1.66	58.2	81.2	14.6
mobilenet_v2_0.35_160	30	1.66	55.7	79.1	10.5
mobilenet_v2_0.35_128	20	1.66	50.8	75.0	6.9
mobilenet_v2_0.35_96	11	1.66	45.5	70.4	4.5

You can download the model from `https://github.com/tensorflow/models/tree/master/research/slim/nets/mobilenet`.

Distortions

We can improve the results by giving tough input images during training. Training images can be generated by cropping, brightening, and deforming the input images randomly. This will help in generating an effective training dataset.

However, there is a disadvantage of enabling distortion here, since bottleneck caching is not useful. Consequently, the input images are not reused, increasing the training time period. There are multiple ways to enable distortion, as shown here:

```
--random_crop
--random_scale
--random_brightness
```

This won't be useful in all cases. For example, it won't be helpful in a digit classifier system, since flipping and distorting the image won't make sense when it comes to producing a possible output.

Hyperparameters

We can try a few more parameters to check whether additional parameters will help to improve results.

Specify them in the form that's given in the following bullet points. The hyperparameters are explained as follows:

- `--learning_rate`: This parameter controls the updates to the final layer while training. If this value is small, the training will take more time. This may not always help when it comes to improving accuracy.

- `--train_batch_size`: This parameter helps with controlling the number of images that are examined during training to estimate the final-layer updates. Once the images are ready, the script splits them into three different sets. The largest set is used in training. This division is mainly useful for preventing the model from recognizing unnecessary patterns in the input images. If a model is trained using a certain background pattern, it won't give a proper result when it faces images with new backgrounds because it remembers unnecessary information from the input images. This is known as **overfitting**.

- `--testing_percentage` and `--validation_percentage` flags: To avoid overfitting, we keep 80% of the data inside the main training set. Of this data, 10% is then used to run validation during the training process and the final 10% is used to test the model.

- `--validation_batch_size`: We can see that the accuracy of validation fluctuates between iterations.

If you are new to this, you can run default values without making any changes to these parameters. Let's jump into building our model. For this, we need the training image data.

Image dataset collection

For our experiment, we need the datasets for cars in good condition as well as in damaged condition. If you have a data source that adheres to the privacy policy, then this is a good place to start. Otherwise, we need to find a way to build our model on top of a dataset. There are multiple datasets that are publicly available. We need to start building our dataset if there is no existing reference of a similar data model because this could be a time-consuming task as well as an important step toward getting better results.

We are going to use a simple Python script to download images from Google. Just make sure that you filter images that can be reused. We don't encourage using pictures with non-reusable licenses.

With the Python script, we will pull out and save the images from Google, and then we will use a library to do the same task. This step is one of the most basic steps for building any ML model.

We will use a Python library called **Beautiful Soup** to scrap images from the internet.

Introduction to Beautiful Soup

Beautiful Soup is a Python library that's used to pull data out of HTML and XML files. It is useful with projects that involve scraping. With this library, we can navigate, search, and modify the HTML and XML files.

This library parses anything you feed in and does tree traversal work on the data. You can ask the library to find all the links whose URLs match `google.com`, find all the links with class bold URLs, or find all the table headers with bold text.

There are a few features that makes it useful, and they are as follows:

1. Beautiful Soup provides us with some simple methods and Pythonic idioms to navigate, search, and modify a parse tree, which is a toolkit that is used to dissect a document and then extract what you need. We need less code to write an application.

2. Beautiful Soup automatically converts incoming documents into Unicode and outgoing documents into UTF-8. Unless the document doesn't specify anything about encoding and Beautiful Soup isn't able to detect any, we don't have to think about encoding. Then, we will have to specify only the original encoding.

3. Beautiful Soup can be used on top of popular Python parsers, such as lxml (https://lxml.de/) and html5lib (https://github.com/html5lib/), and lets you try various parsing strategies or trade speed for flexibility.

4. Beautiful Soup saves you time by extracting the information you need and so makes your job easier.

Here is the simple version of the code:

```
import argparse
import json
import itertools
import logging
import re
import os
import uuid
import sys
from urllib.request import urlopen, Request
from bs4 import BeautifulSoup
#logger will be useful for your debugging need
def configure_logging():
logger = logging.getLogger()
logger.setLevel(logging.DEBUG)
handler = logging.StreamHandler()
handler.setFormatter(
logging.Formatter('[%(asctime)s %(levelname)s %(module)s]: %(message)s'))
logger.addHandler(handler)
return logger
logger = configure_logging()
```

Setting the user-agent to avoid 403 error code:

```
REQUEST_HEADER = {
'User-Agent': "Mozilla/5.0 (Windows NT 6.1; WOW64) AppleWebKit/537.36
(KHTML, like Gecko) Chrome/43.0.2357.134 Safari/537.36"}
def get_soup(url, header):
```

```
response = urlopen(Request(url, headers=header))
return BeautifulSoup(response, 'html.parser')
# initialize place for links
def get_query_url(query):
return "https://www.google.co.in/search?q=%ssource=lnmstbm=isch" % query
# pull out specific data through navigating into source data tree
def extract_images_from_soup(soup):
image_elements = soup.find_all("div", {"class": "rg_meta"})
metadata_dicts = (json.loads(e.text) for e in image_elements)
link_type_records = ((d["ou"], d["ity"]) for d in metadata_dicts)
return link_type_records
```

Pass the number of images you want to pull out. By default google provides 100 images:

```
def extract_images(query, num_images):
url = get_query_url(query)
logger.info("Souping")
soup = get_soup(url, REQUEST_HEADER)
logger.info("Extracting image urls")
link_type_records = extract_images_from_soup(soup)
return itertools.islice(link_type_records, num_images)
def get_raw_image(url):
req = Request(url, headers=REQUEST_HEADER)
resp = urlopen(req)
return resp.read()
```

Saving all the downloaded images along with its extension, as shown in the following code block:

```
def save_image(raw_image, image_type, save_directory):
extension = image_type if image_type else 'jpg'
file_name = str(uuid.uuid4().hex) + "." + extension
save_path = os.path.join(save_directory, file_name)
with open(save_path, 'wb+') as image_file:
image_file.write(raw_image)
def download_images_to_dir(images, save_directory, num_images):
for i, (url, image_type) in enumerate(images):
try:
logger.info("Making request (%d/%d): %s", i, num_images, url)
raw_image = get_raw_image(url)
save_image(raw_image, image_type, save_directory)
except Exception as e:
logger.exception(e)
def run(query, save_directory, num_images=100):
query = '+'.join(query.split())
logger.info("Extracting image links")
images = extract_images(query, num_images)
logger.info("Downloading images")
```

```
download_images_to_dir(images, save_directory, num_images)
logger.info("Finished")
#main method to initiate the scrapper
def main():
parser = argparse.ArgumentParser(description='Scrape Google images')
#change the search term here
parser.add_argument('-s', '--search', default='apple', type=str,
help='search term')
```

Change number of images parameter here. By default it is set to 1, as shown in following code:

```
parser.add_argument('-n', '--num_images', default=1, type=int, help='num
images to save')
#change path according to your need
parser.add_argument('-d', '--directory',
default='/Users/karthikeyan/Downloads/', type=str, help='save directory')
args = parser.parse_args()
run(args.search, args.directory, args.num_images)
if __name__ == '__main__':
main()
```

Save the script as a Python file and then run the code by executing the following command:

```
python imageScrapper.py --search "alien" --num_images 10 --directory
"/Users/Karthikeyan/Downloads"
```

Google image scraping with a better library, including more configurable options. We will use https://github.com/hardikvasa/google-images-download.

This is a command line Python program that's used to search for keywords or key phrases on Google Images, and optionally download images to your computer. You can also invoke this script from another Python file.

This is a small and ready-to-run program. No dependencies are required for it to be installed if you only want to download up to 100 images per keyword. If you want more than 100 images per keyword, then you will need to install the Selenium library, along with **ChromeDriver**. Detailed instructions are provided in the *Troubleshooting* section.

You can use a library with more useful options.

If you prefer command line-based installation, use the following code:

```
$ git clone https://github.com/hardikvasa/google-images-download.git
$ cd google-images-download && sudo python setup.py install
```

Alternatively, you can install the library through `pip`:

```
$ pip install google_images_download
```

If installed via `pip` or using a **command language interpreter** (**CLI**), use the following command:

```
$ googleimagesdownload [Arguments...]
```

If downloaded via the UI from `github.com`, unzip the downloaded file, go to the `google_images_download` directory, and use one of the following commands:

```
$ python3 google_images_download.py [Arguments...]
```

```
$ python google_images_download.py [Arguments...]
```

If you want to use this library from another Python file, use this command:

```
from google_images_download import google_images_download
response = google_images_download.googleimagesdownload()
 absolute_image_paths = response.download({<Arguments...>})
```

You can either pass the arguments directly from the command, as shown in the following examples, or you can pass it through a config file.

You can pass more than one record through a config file. The following sample consists of two sets of records. The code will iterate through each of the records and download images based on the arguments that are passed.

The following is a sample of what a config file looks like:

```
{
 "Records": [
 {
 "keywords": "apple",
 "limit": 55,
 "color": "red",
 "print_urls": true
 },
 {
 "keywords": "oranges",
 "limit": 105,
 "size": "large",
 "print_urls": true
 }
 ]
 }
```

Examples

If you are calling this library from another Python file, the following is the sample code from Google:

```
_images_download import google_images_download

#importing the library

response = google_images_download.googleimagesdownload()

#class instantiation

arguments = {"keywords":"apple, beach, cat","limit":30,"print_urls":True}
#creating list of arguments
paths = response.download(arguments) #passing the arguments to the function
print(paths)

#printing absolute paths of the downloaded images
```

If you are passing arguments from a config file, simply pass the `config_file` argument with the name of your **JSON** file:

```
$ googleimagesdownload -cf example.json
```

The following is a simple example of using keywords and limit arguments:

```
$ googleimagesdownload --keywords "apple, beach, cat" --limit 20
```

Using suffix keywords allows you to specify words after the main keyword. For example, if the keyword is `car` and the suffix keywords are `red` and `blue`, then it will first search for a red car and then a blue car:

```
$ googleimagesdownload --k "car" -sk 'yellow,blue,green' -l 10
```

To use the short-hand command, use the following code:

```
$ googleimagesdownload -k "apple, beach, cat" -l 20
```

To download images with specific image extension, or formats, use the following code:

```
$ googleimagesdownload --keywords "logo" --format svg
```

To use color filters for the images, use the following code:

```
$ googleimagesdownload -k "playground" -l 20 -co red
```

To use non-English keywords for image searches, use the following code:

```
$ googleimagesdownload -k "北极熊" -l 5
```

To download images from the Google Images link, use the following code:

```
$ googleimagesdownload -k "sample" -u <google images page URL>
```

To save images in a specific main directory (instead of in Downloads), use the following code:

```
$ googleimagesdownload -k "boat" -o "boat_new"
```

To download one single image within the image URL, use the following code:

```
$ googleimagesdownload --keywords "baloons" --single_image <URL of the images>
```

To download images with size and type constraints, use the following code:

```
$ googleimagesdownload --keywords "baloons" --size medium --type animated
```

To download images with specific usage rights, use the following code:

```
$ googleimagesdownload --keywords "universe" --usage_rights labeled-for-reuse
```

To download images with specific color types, use the following code:

```
$ googleimagesdownload --keywords "flowers" --color_type black-and-white
```

To download images with specific aspect ratios, use the following code:

```
$ googleimagesdownload --keywords "universe" --aspect_ratio panoramic
```

To download images that are similar to the image in the image URL that you provided (known as a reverse image search), use the following code:

```
$ googleimagesdownload -si <image url> -l 10
```

To download images from a specific website or domain name for a given keyword, use the following code:

```
$ googleimagesdownload --keywords "universe" --specific_site google.com
```

The images will be downloaded to their own sub-directories inside the main directory (either the one you provided or in `Downloads`) in the same folder you are in.

Now, we need to start preparing our dataset.

Dataset preparation

We need to build four different datasets. For car damage detection, we will think about all the possible inputs. It could be a car in good condition, or a car with different damage levels, or it could be an unrelated image of a car.

We will do the same as shown in the following screenshots:

```
[Karthikeyans-MacBook-Pro:google-images-download karthikeyan$ googleimagesdownload -k "car" -sk 'red,blue' -l 50

Item no.: 1 --> Item name = car red
Evaluating...
Starting Download...
Completed Image ===> 1. 71jrcuoujbl._sl1500_.jpg
Completed Image ===> 2. lamborghini_huracan_slideshow_lead.jpg
Completed Image ===> 3. item_xl_22318852_30176327.jpg
Completed Image ===> 4. car4-1196370.jpg
Completed Image ===> 5. 598cc71515000084208b6139.jpg
Completed Image ===> 6. image_3568d179-cb57-4232-9572-034f3c302dcd_1024x1024.jpg
Completed Image ===> 7. maxresdefault.jpg
Completed Image ===> 8. 20170309-red-cars-at-geneva-vlad-savov11.0.jpg
Completed Image ===> 9. 61rmtid79wl._sx425_.jpg
Completed Image ===> 10. 51pja9m4xdl.jpg
Completed Image ===> 11. ferrari+f12+berlinetta.jpg
Completed Image ===> 12. ferrari-laferrari-12v-ride-on-car-red-a4d.jpg
Completed Image ===> 13. p_red_car.jpg
Completed Image ===> 14. big-new-bobby-car-red-800056200_00.jpeg
Completed Image ===> 15. 71ok%2bxwygfl._sx425_.jpg
Completed Image ===> 16. dsc_0001__80292.1536633226.1280.1280.jpg
Completed Image ===> 17. ferrari-458-super-car-big-remote-control-car-red-sunshine-original-imaeuz56fcauwjzk.jpeg
Completed Image ===> 18. 919ec5a_1433856.jpg
Completed Image ===> 19. gqvhhhx.jpg
Completed Image ===> 20. gla_classic.jpeg
Completed Image ===> 21. aston-martin-vanquish-s-raf-red-arrows-829649.jpg
Completed Image ===> 22. maxresdefault.jpg
Completed Image ===> 23. 995-panda-original-imaee4za5fwpfa4k.jpeg
Completed Image ===> 24. car-30984_960_720.png
Completed Image ===> 25. 524c28b2-d349-495f-81e0-eec7442ac28f_1.6bd30cb6b69ad4327d9102a4f07ded51.jpeg
```

Here is the dataset to identify heavily damaged cars:

```
googleimagesdownload -k "heavily damaged car" -sk
'red,blue,white,black,green,brown,pink,yellow' -l 500
```

Here are some sample pictures that were captured for heavily damaged cars that are red:

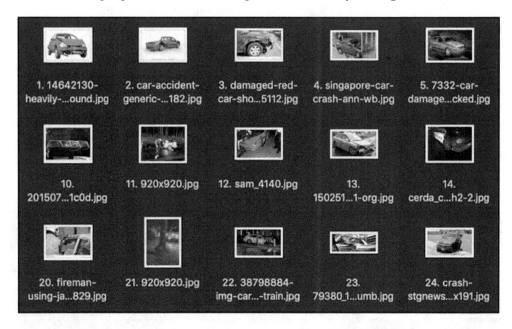

Here are some sample pictures that were captured for heavily damaged cars that are blue:

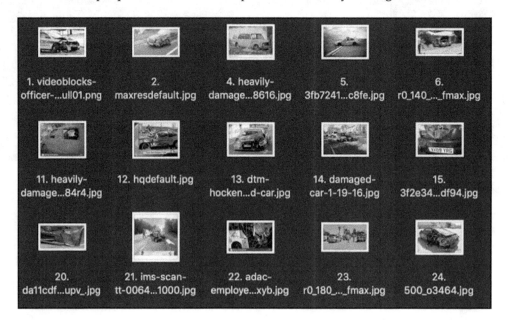

We also have another set of images for cars with slightly less damage:

```
googleimagesdownload -k "car dent" -sk
'red,blue,white,black,green,brown,pink,yellow' -l 500
```

Here are some sample pictures that were captured for dented cars that are red:

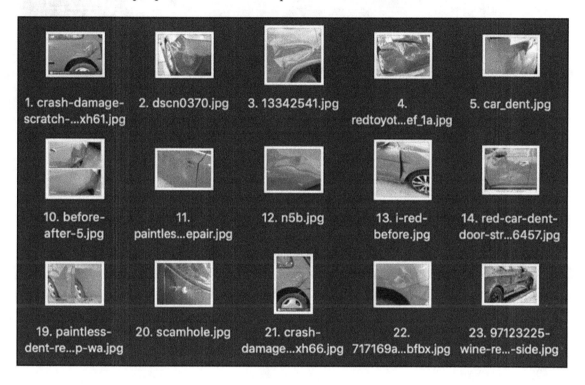

Here are some sample pictures that were captured for dented cars that are blue:

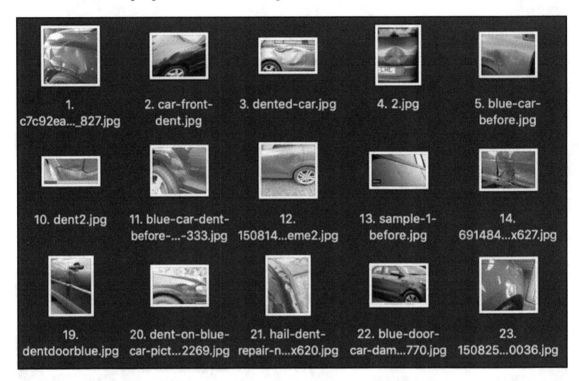

The following command can be used to retrieve a dataset for normal cars without any damage applied to them:

```
googleimagesdownload -k "car" -l 500
```

Here are some sample pictures that have been captured for cars that are red:

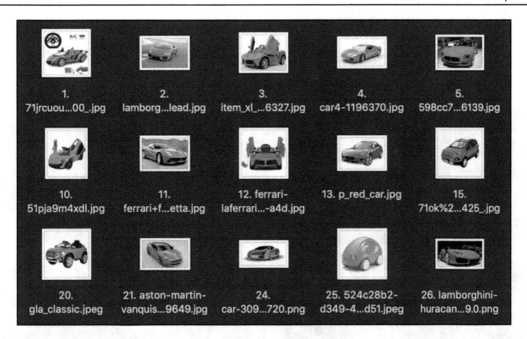

Here are some sample pictures that have been captured for cars that are blue:

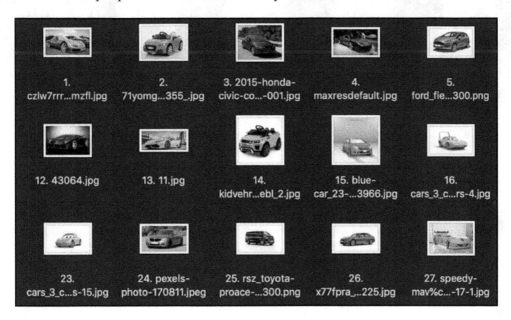

The following command can be used to retrieve random objects that aren't cars:

```
googleimagesdownload -k "bike,flight,home,road,tv" -l 500
```

Here are some sample pictures that have been captured for bikes:

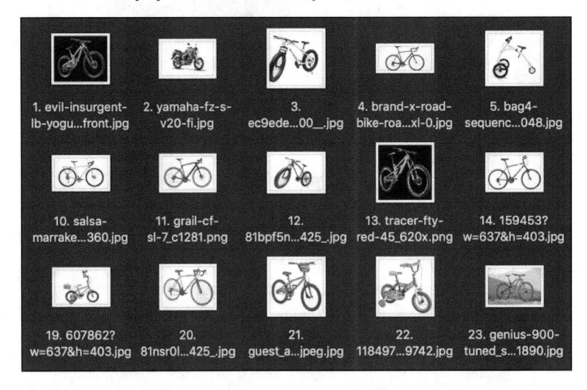

Here are some sample pictures that have been captured for flights:

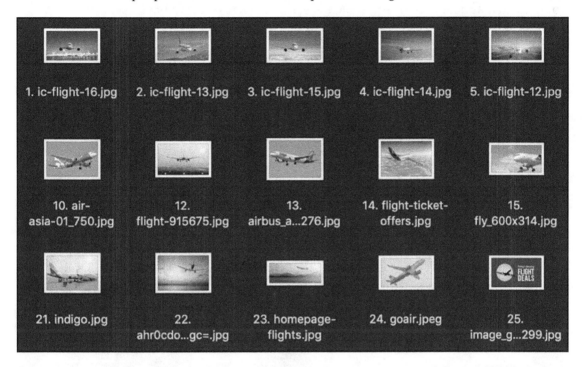

Once we have 500 images for each dataset, it's time to do some training. In ideal conditions, we should have at least 1,000 images for each dataset.

The main issue we will face here is in removing the noise data. For our example, we are going to do that manually. There are a few sample images that we have listed here that could be noise, and don't provide valid input so that we can build the data model:

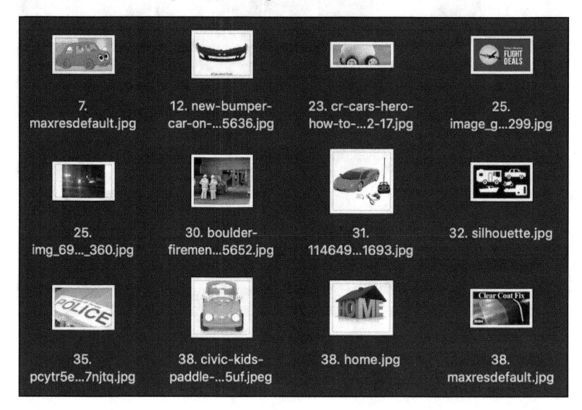

Once we have all the image datasets ready, we can move on to our top four categories. Right now, all the images are separated by colors and categories, as shown in the following screenshot:

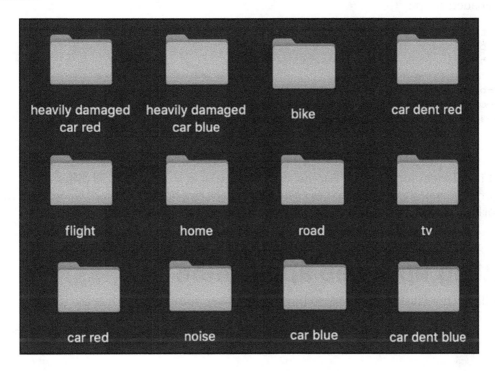

We will group them into damaged car, car with dent, car, and not a car:

Running the training script

With all the parameter-related details discussed, we can start the training with the downloaded script:

```
python retrain.py \
--bottleneck_dir=./ \
--how_many_training_steps=4000 \
--model_dir=./ \
--output_graph=./retrained_graph.pb \
--output_labels=retrained_labels.txt \
--architecture=" mobilenet_v2_1.4_224" \
--image_dir=/Users/karthikeyan/Documents/ /book/Chapter5/images
```

Based on our processor's capability, as well as the number of images we have, the script might take longer for training. For me, it took more than 10 hours for 50 different car categories containing 10,000 images each. Once the script has completed, we will get the TensorFlow model in its output.

Setting up a web application

We will use the **Flask** framework to build a simple application to detect the car's damage.

 To learn more about Flask, please refer to https://www.fullstackpython.com/flask.html.

We are not going to go deeper into Flask basics here. Instead, we are simply adding our model with an existing file upload example from Flask.

The file's structure is shown in the following screenshot:

Here is a list of the contents in `app.py`:

```
import os
import glob
from classify import prediction
import tensorflow as tf
import thread
import time
from flask import Flask, render_template, request, redirect, url_for,
send_from_directory,flash
from werkzeug import secure_filename
app = Flask(__name__)
app.config['UPLOAD_FOLDER'] = 'uploads/'
app.config['ALLOWED_EXTENSIONS'] = set(['jpg', 'jpeg'])
app.config['SECRET_KEY'] = '7d441f27d441f27567d441f2b6176a'
def allowed_file(filename):
return '.' in filename and \
filename.rsplit('.', 1)[1] in app.config['ALLOWED_EXTENSIONS']
@app.route('/')
```

```
def index():
return render_template('index.html')
@app.route('/upload', methods=['POST'])
def upload():
file = request.files['file']
if file and allowed_file(file.filename):
filename = secure_filename(file.filename)
filename = str(len(os.listdir(app.config['UPLOAD_FOLDER']))+1)+'.jpg'
file_name_full_path = os.path.join(app.config['UPLOAD_FOLDER'], filename)
file.save(file_name_full_path)
return render_template('upload_success.html')
@app.route('/uploads/<filename>')
def uploaded_file(filename):
return send_from_directory(app.config['UPLOAD_FOLDER'],
filename)
@app.route('/claim', methods=['POST'])
def predict():
list_of_files =
glob.glob('/Users/karthikeyan/Documents/code/play/acko/cardamage/Car-
Damage-Detector/uploads/*.jpg') # * means all if need specific format then
*.csv
latest_file = max(list_of_files, key=os.path.getctime)
print(latest_file)
image_path = latest_file
```

Next code block helps us with printing the output:

```
#print(max(glob.glob(r'uploads\*.jpg'), key=os.path.getmtime))
with tf.Graph().as_default():
human_string, score= prediction(image_path)
print('model one value' + str(human_string))
print('model one value' + str(score))
if (human_string == 'car'):
label_text = 'This is not a damaged car with confidence ' + str(score) +
'%. Please upload a damaged car image'
print(image_path)
return render_template('front.html', text = label_text,
filename="http://localhost:5000/uploads/"+os.path.basename(image_path))
elif (human_string == 'low'):
label_text = 'This is a low damaged car with '+ str(score) + '%
confidence.'
print(image_path)
```

After printing the image path, go through the following code:

```
return render_template('front.html', text = label_text,
filename="http://localhost:5000/uploads/"+os.path.basename(image_path))
elif (human_string == 'high'):
label_text = 'This is a high damaged car with '+ str(score) + '%
confidence.'
print(image_path)
return render_template('front.html', text = label_text,
filename="http://localhost:5000/uploads/"+os.path.basename(image_path))
elif (human_string == 'not'):
label_text = 'This is not the image of a car with confidence ' + str(score)
+ '%. Please upload the car image.'
print(image_path)
return render_template('front.html', text = label_text,
filename="http://localhost:5000/uploads/"+os.path.basename(image_path))
def cleanDirectory(threadName,delay):
```

The while loop starts from here:

```
while True:
time.sleep(delay)
print ("Cleaning Up Directory")
filelist = [ f for f in (os.listdir(app.config['UPLOAD_FOLDER'])) ]
for f in filelist:
#os.remove("Uploads/"+f)
os.remove(os.path.join(app.config['UPLOAD_FOLDER'], f))
if __name__ == '__main__':
try:
_thread.start_new_thread( cleanDirectory, ("Cleaning Thread", 99999999, ) )
except:
print("Error: unable to start thread" )
app.run()
Classify.py does the model classification using TensorFlow.
import tensorflow as tf
import sys
import os
import urllib
```

Disable TensorFlow compilation warnings:

```
os.environ['TF_CPP_MIN_LOG_LEVEL']='2'
import tensorflow as tf
def prediction(image_path):
image_data = tf.gfile.FastGFile(image_path, 'rb').read()
print(image_path)
label_lines = [line.rstrip() for line
in tf.gfile.GFile(r"./models/tf_files/retrained_labels.txt")]
```

```
with tf.gfile.FastGFile(r"./models/tf_files/retrained_graph.pb", 'rb') as
f:
graph_def = tf.GraphDef()
graph_def.ParseFromString(f.read())
_ = tf.import_graph_def(graph_def, name='')
with tf.Session() as sess:
```

Once `image_data` is given as the input to the graph, we then receive the first prediction:

```
softmax_tensor = sess.graph.get_tensor_by_name('final_result:0')
predictions = sess.run(softmax_tensor, \
{'DecodeJpeg/contents:0': image_data})
top_k = predictions[0].argsort()[-len(predictions[0]):][::-1]
for node_id in top_k:
count = 1
human_string = label_lines[node_id]
score = predictions[0][node_id]
print(count)
count += 1
print('%s (score = %.5f)' % (human_string, score))
score = (round((score * 100), 2))
return human_string, score
```

The controller Python files are lined in frontend HTML files:

```
<!DOCTYPE html>
<html lang="en">
<head>
<meta charset="utf-8">
<meta name="viewport" content="width=device-width, initial-scale=1,
shrink-to-fit=no">
<meta name="description" content="">
<meta name="author" content="Karthikeyan NG">
<title>Damage Estimator</title>
<!-- Bootstrap core CSS -->
<link href="{{ url_for('static',
filename='vendor/bootstrap/css/bootstrap.min.css') }}" rel="stylesheet"/>
<!-- Custom fonts for this template -->
<link href="{{ url_for('static', filename='vendor/font-awesome/css/font-
awesome.min.css') }}" rel="stylesheet" type="text/css"/>
<link
href='https://fonts.googleapis.com/css?family=Open+Sans:300italic,400italic
,600italic,700italic,800italic,400,300,600,700,800' rel='stylesheet'
type='text/css'>
<link
href='https://fonts.googleapis.com/css?family=Merriweather:400,300,300itali
c,400italic,700,700italic,900,900italic' rel='stylesheet' type='text/css'>
<!-- Plugin CSS -->
```

```
<link href="{{ url_for('static', filename='vendor/magnific-popup/magnific-
popup.css') }}" rel="stylesheet" />
<!-- Custom styles for this template -->
<link href="{{ url_for('static', filename='css/creative.min.css') }}"
rel="stylesheet" />
</head>
<body id="page-top">
<!-- Navigation -->
<nav class="navbar navbar-expand-lg navbar-light fixed-top" id="mainNav">
<a class="navbar-brand" href="#page-top">Damage Estimator</a>
<button class="navbar-toggler navbar-toggler-right" type="button" data-
toggle="collapse" data-target="#navbarResponsive" aria-
controls="navbarResponsive" aria-expanded="false" aria-label="Toggle
navigation">
<span class="navbar-toggler-icon"></span>
</button>
<div class="collapse navbar-collapse" id="navbarResponsive">
</div>
</nav>
<section class="bg-primary" id="about">
<div class="container">
<div class="row">
<div class="col-lg-8 mx-auto text-center">
<h2 class="section-heading text-white">Do you have a damaged vehicle?</h2>
<hr class="light">
<p class="text-faded">Machine Learning allows for a classification process
that is automated and makes lesser error. Besides risk group
classification, Deep Learning algorithms can be applied to images of
vehicle damage, allowing for automated claim classification.</p>
<br/>
<div class="contr"><h4 class="section-heading text-white">Select the file
(image) and Upload</h4></div>
<br/>
<form action="upload" method="post" enctype="multipart/form-data">
<div class="form-group">
<input type="file" name="file" class="file">
<div class="input-group col-xs-12">
<span class="input-group-addon"><i class="glyphicon glyphicon-
picture"></i></span>
<input type="text" class="form-control input-lg" disabled
placeholder="Upload Image">
<span class="input-group-btn">
<button class="browse btn btn-primary input-lg" type="button"><i
class="glyphicon glyphicon-search"></i> Browse</button>
</span>
</div>
</div>
<input type="submit" class="btn btn-primary" value="Upload"><br /><br />
```

```
          </form>
        </div>
      </div>
    </section>
```

In continuation with the previous script, let's Bootstrap the core JavaScript:

```
    <!-- Bootstrap core JavaScript -->
    <script src="{{ url_for('static', filename='vendor/jquery/jquery.min.js')
}}"></script>
    <script src="{{ url_for('static', filename='vendor/popper/popper.min.js')
}}"></script>
    <script src="{{ url_for('static',
filename='vendor/bootstrap/js/bootstrap.min.js') }}"></script>
    <!-- Plugin JavaScript -->
    <script src="{{ url_for('static', filename='vendor/jquery-
easing/jquery.easing.min.js') }}"></script>
    <script src="{{ url_for('static',
filename='vendor/scrollreveal/scrollreveal.min.js') }}"></script>
    <script src="{{ url_for('static', filename='vendor/magnific-
popup/jquery.magnific-popup.min.js') }}"></script>
    <!-- Custom scripts for this template -->
    <script src="{{ url_for('static', filename='js/creative.min.js')
}}"></script>
    <script>
    $(document).on('click', '.browse', function(){
    var file = $(this).parent().parent().parent().find('.file');
    file.trigger('click');
    });
    $(document).on('change', '.file', function(){
    $(this).parent().find('.form-
control').val($(this).val().replace(/C:\\fakepath\\/i, ''));
    });
    </script>
    </body>
    </html>
```

You can pull the rest of the file's content directly from the GitHub repository. Once the complete file structure is ready, you can run the application from the command line, as follows:

```
$ python app.py
```

Now, launch your browser with `http://localhost:5000/`:

```
Karthikeyans-MacBook-Pro:Car-Damage-Detector karthikeyan$ python app.py
Error: unable to start thread
 * Running on http://127.0.0.1:5000/ (Press CTRL+C to quit)
127.0.0.1 - - [19/Feb/2019 23:28:35] "GET / HTTP/1.1" 200 -
127.0.0.1 - - [19/Feb/2019 23:28:35] "GET /static/vendor/bootstrap/css/bootstrap.min.css HTTP/1.1" 200 -
127.0.0.1 - - [19/Feb/2019 23:28:35] "GET /static/vendor/font-awesome/css/font-awesome.min.css HTTP/1.1"
 200 -
127.0.0.1 - - [19/Feb/2019 23:28:35] "GET /static/vendor/magnific-popup/magnific-popup.css HTTP/1.1" 200
 -
127.0.0.1 - - [19/Feb/2019 23:28:35] "GET /static/css/creative.min.css HTTP/1.1" 200 -
127.0.0.1 - - [19/Feb/2019 23:28:35] "GET /static/vendor/jquery/jquery.min.js HTTP/1.1" 200 -
127.0.0.1 - - [19/Feb/2019 23:28:35] "GET /static/vendor/popper/popper.min.js HTTP/1.1" 200 -
127.0.0.1 - - [19/Feb/2019 23:28:35] "GET /static/vendor/bootstrap/js/bootstrap.min.js HTTP/1.1" 200 -
127.0.0.1 - - [19/Feb/2019 23:28:35] "GET /static/vendor/jquery-easing/jquery.easing.min.js HTTP/1.1" 20
0 -
127.0.0.1 - - [19/Feb/2019 23:28:35] "GET /static/vendor/scrollreveal/scrollreveal.min.js HTTP/1.1" 200
 -
127.0.0.1 - - [19/Feb/2019 23:28:35] "GET /static/vendor/magnific-popup/jquery.magnific-popup.min.js HTT
P/1.1" 200 -
127.0.0.1 - - [19/Feb/2019 23:28:35] "GET /static/js/creative.min.js HTTP/1.1" 200 -
127.0.0.1 - - [19/Feb/2019 23:28:36] "GET /favicon.ico HTTP/1.1" 404 -
127.0.0.1 - - [19/Feb/2019 23:30:08] "POST /upload HTTP/1.1" 200 -
/Users/karthikeyan/Documents/code/play/acko/cardamage/Car-Damage-Detector/uploads/19.jpg
/Users/karthikeyan/Documents/code/play/acko/cardamage/Car-Damage-Detector/uploads/19.jpg
1
low (score = 0.49069)
model one valuelow
model one value49.07
/Users/karthikeyan/Documents/code/play/acko/cardamage/Car-Damage-Detector/uploads/19.jpg
127.0.0.1 - - [19/Feb/2019 23:30:26] "POST /claim HTTP/1.1" 200 -
127.0.0.1 - - [19/Feb/2019 23:30:54] "GET /uploads/19.jpg HTTP/1.1" 200 -
```

The following are a few screenshots from the application.

Here is the home page after running the application:

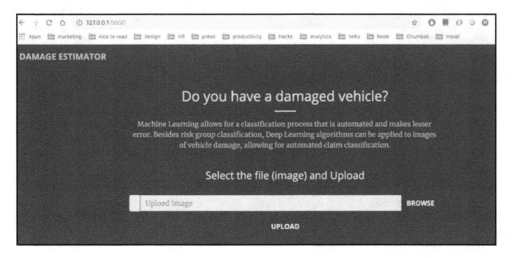

Here is the screen after uploading the image:

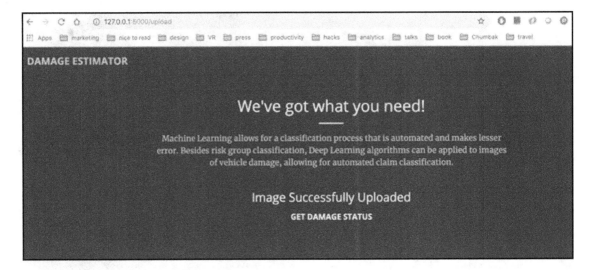

Here is a screenshot showing an image of a car with low level damage

The data in the preceding screenshot may not be accurate, since our dataset size is very small.

The following is a screenshot showing an image of the car prediction model that does not show a car:

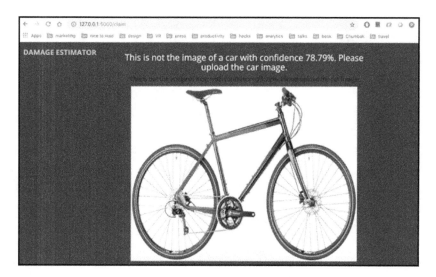

Summary

In this chapter, we have seen how we can build a model from scratch and train it using TensorFlow.

With this knowledge, we can start building more Android and iOS-based applications in the upcoming chapters.

PyTorch Experiments on NLP and RNN

6

In this chapter, we are going to deep dive into the PyTorch library on **natural language processing** (**NLP**) and other experiments. Then, we will convert the developed model into a format that can be used in an Android or iOS application using TensorFlow and CoreML.

In this chapter, we will cover the following topics:

- Introduction to PyTorch features and installation
- Using variables in PyTorch
- Building our own model network
- Classifying **recurrent neural networks** (**RNN**)
- Natural language processing

PyTorch

PyTorch is a Python-based library that's used to perform scientific computing operations with GPUs. It helps by performing faster experimentation to run production-grade ecosystems and distribute the training of libraries. It also provides two high-level features: tensor computations and building neural networks on tape-based autograd systems.

The features of PyTorch

PyTorch provides an end-to-end deep learning system. Its features are as follows:

- **Python utilization**: PyTorch is not simply a Python binding to a C++ framework. It is deeply integrated in Python so that it can be used with other popular libraries and frameworks.

- **Tools and libraries**: It has an active community of researchers and developers in the areas of computer vision and reinforcement learning.
- **Flexible frontend**: This includes ease of use and hybrid in eager mode, accelerate speeds and seamless transitions to graph mode, and functionality and optimization in C++ runtime.
- **Cloud support**: This is supported on all of the major cloud platforms, allowing for seamless development and scaling with prebuilt images, so that it is able to run as a production-grade application.
- **Distributed training**: This includes performance optimization with the advantage of native support for the asynchronous execution of operations and peer-to-peer (p2p) communications, so that we can access both C++ and Python.
- **Native support for ONNX**: We can export models into the standard **Open Neural Network Exchange** (**ONNX**) format for access to other platforms, runtimes, and visualizers.

Installing PyTorch

There is a stable version of PyTorch available at the time of writing this book, that is, 1.0. There is also a nightly preview build available if you want to have a hands-on look at the latest code repository. You need to have the dependencies installed based on your package manager. **Anaconda** is the recommended package manager, and it installs all the dependencies automatically. **LibTorch** is only available for C++. Here is a grid showing the installation options that are available for installing PyTorch:

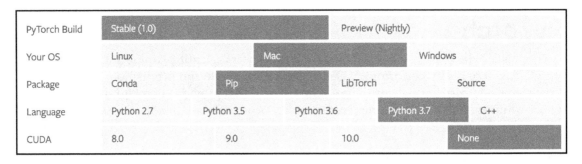

PyTorch Build	Stable (1.0)			Preview (Nightly)	
Your OS	Linux		Mac		Windows
Package	Conda	Pip		LibTorch	Source
Language	Python 2.7	Python 3.5	Python 3.6	Python 3.7	C++
CUDA	8.0	9.0		10.0	None

The preceding screenshot specifies the package grid that was used while this book was being written. You can pick any package grid as per your hardware configuration availability.

To install PyTorch, and to start Jupyter Notebook, run the following command:

```
python --version
sudo brew install python3
brew install python3
pip3 install --upgrade pip
pip3 install jupyter
jupyter notebook
```

The installation of PyTorch is shown in the following screenshot:

```
Karthikeyans-MacBook-Pro:code karthikeyan$ pip3 install jupyter
Collecting jupyter
  Using cached https://files.pythonhosted.org/packages/83/df/0f5dd132200728a86190397e1ea87cd76244e42d39e
c5e88efd25b2abd7e/jupyter-1.0.0-py2.py3-none-any.whl
Collecting ipywidgets (from jupyter)
  Using cached https://files.pythonhosted.org/packages/30/9a/a008c7b1183fac9e52066d80a379b3c64eab535bd9d
86cdc29a0b766fd82/ipywidgets-7.4.2-py2.py3-none-any.whl
Collecting jupyter-console (from jupyter)
  Downloading https://files.pythonhosted.org/packages/cb/ee/6374ae8c21b7d0847f9c3722dcdfac986b8e54fa9ad9
ea66e1eb6320d2b8/jupyter_console-6.0.0-py2.py3-none-any.whl
Collecting nbconvert (from jupyter)
  Using cached https://files.pythonhosted.org/packages/b8/39/1e67fea74dc9577cc49f9863fe3ec824e525d1304ab
6027d95a94cd586f5/nbconvert-5.4.1-py2.py3-none-any.whl
Collecting ipykernel (from jupyter)
  Downloading https://files.pythonhosted.org/packages/d8/b0/f0be5c5ab335196f5cce96e5b889a4fcf5bfe462eb0a
cc05cd7e2caf65eb/ipykernel-5.1.0-py3-none-any.whl (113kB)
    100% |████████████████████████████████| 122kB 1.7MB/s
Collecting qtconsole (from jupyter)
  Using cached https://files.pythonhosted.org/packages/e0/7a/8aefbc0ed078dec7951ac9a06dcd1869243ecd7bcbc
e26fa47bf5e469a8f/qtconsole-4.4.3-py2.py3-none-any.whl
Collecting notebook (from jupyter)
  Using cached https://files.pythonhosted.org/packages/0a/d8/4e9521354ed3d730ba6d8a5af440b66c73245ef46be
706e51bead71afc21/notebook-5.7.6-py2.py3-none-any.whl
Collecting widgetsnbextension~=3.4.0 (from ipywidgets->jupyter)
  Using cached https://files.pythonhosted.org/packages/8a/81/35789a3952afb48238289171728072d26d6e76649dd
c8b3588657a2d78c1/widgetsnbextension-3.4.2-py2.py3-none-any.whl
Collecting nbformat>=4.2.0 (from ipywidgets->jupyter)
  Using cached https://files.pythonhosted.org/packages/da/27/9a654d2b6cc1eaa517d1c5a4405166c7f6d72f04f6e
7eea41855fe808a46/nbformat-4.4.0-py2.py3-none-any.whl
Collecting ipython>=4.0.0; python_version >= "3.3" (from ipywidgets->jupyter)
  Downloading https://files.pythonhosted.org/packages/14/3b/3fcf422a99a04ee493e6a4fc3014e3c8ff484a7feed2
38fef68bdc285085/ipython-7.3.0-py3-none-any.whl (768kB)
    100% |████████████████████████████████| 778kB 8.0MB/s
Collecting traitlets>=4.3.1 (from ipywidgets->jupyter)
```

When you initiate the Jupyter Notebook, a new browser session opens up with an empty notebook, as shown here:

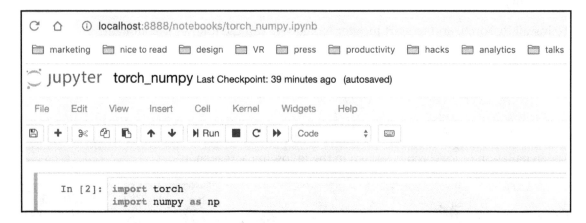

Let's look at the basics of PyTorch.

PyTorch basics

Now that PyTorch has been installed, we can start experimenting with it. We will start with torch and numpy.

From the top menu, create a new notebook and include the following code:

```
# first basic understanding on PyTorch
# book: AI for Mobile application projects

import torch
import numpy as np

# convert numpy to tensor or vise versa
numpy_data = np.arange(8).reshape((2, 4))
torch_data = torch.from_numpy(numpy_data)
#convert tensor to array
tensor2array = torch_data.numpy()

#Print the results
print
(
 '\nnumpy array:', numpy_data,      # [[0 1 2 3], [4 5 6 7]]
 '\ntorch tensor:', torch_data,     # 0 1 2 3\n 4 5 6 7 [torch.LongTensor of
size 2x3]
```

```
 '\ntensor to array:', tensor2array, # [[0 1 2 3], [4 5 6 7]]
)
```

Now, let's do some mathematical operations:

```
# abs method on numpy
numpy_data = [-1, -2, 1, 2]
tensor = torch.FloatTensor(numpy_data) # 32-bit floating point

#print the results
print
(
 '\nabs',
 '\nnumpy: ', np.abs(numpy_data), # [1 2 1 2]
 '\ntorch: ', torch.abs(tensor) # [1 2 1 2]
)

# sin method on numpy
#print the results
print
(
 '\nsin',
 '\nnumpy: ', np.sin(numpy_data), # [-0.84147098 -0.90929743 0.84147098
0.90929743]
 '\ntorch: ', torch.sin(tensor) # [-0.8415 -0.9093 0.8415 0.9093]
)
```

Let's calculate the mean method and print the results:

```
#print the results
print
(
 '\nmean',
 '\nnumpy: ', np.mean(data), # 0.0
 '\ntorch: ', torch.mean(tensor) # 0.0
)

# matrix multiplication with numpy
numpy_data = [[1,2], [3,4]]
tensor = torch.FloatTensor(numpy_data) # 32-bit floating point
# correct method and print the results
print(
 '\nmatrix multiplication (matmul)',
 '\nnumpy: ', np.matmul(numpy_data, numpy_data), # [[7, 10], [15, 22]]
 '\ntorch: ', torch.mm(tensor, tensor) # [[7, 10], [15, 22]]
)
```

The following code shows the output of the mathematical operations:

```
numpy array: [[0 1 2 3]
 [4 5 6 7]]
torch tensor: tensor([[0, 1, 2, 3],
        [4, 5, 6, 7]])
tensor to array: [[0 1 2 3]
 [4 5 6 7]]

abs
numpy:  [1 2 1 2]
torch:  tensor([1., 2., 1., 2.])

sin
numpy:  [-0.84147098 -0.90929743  0.84147098  0.90929743]
torch:  tensor([-0.8415, -0.9093,  0.8415,  0.9093])

mean
numpy:  0.0
torch:  tensor(0.)

matrix multiplication (matmul)
numpy:  [[ 7 10]
 [15 22]]
torch:  tensor([[ 7., 10.],
        [15., 22.]])
```

Now, let's look at how to use different variables in PyTorch.

Using variables in PyTorch

Variables in `torch` are used to build a computational graph. Whenever a variable is calculated, it builds a computational graph. This computational graph is used to connect all the calculation steps (nodes), and finally, when the error is reversed, the modification range (gradient) in all the variables is calculated at once. In comparison, `tensor` does not have this ability. We will look into this difference with a simple example:

```
import torch
from torch.autograd import Variable

# Variable in torch is to build a computational graph,
# So torch does not have placeholder, torch can just pass variable to the
computational graph.

tensor = torch.FloatTensor([[1,2,3],[4,5,6]]) # build a tensor
variable = Variable(tensor, requires_grad=True) # build a variable, usually
```

```
for compute gradients

print(tensor) # [torch.FloatTensor of size 2x3]
print(variable) # [torch.FloatTensor of size 2x3]

# till now the tensor and variable looks similar.
# However, the variable is a part of the graph, it's a part of the auto-
gradient.

#Now we will calculate the mean value on tensor(X^2)
t_out = torch.mean(tensor*tensor)

#Now we will calculate the mean value on variable(X^2)
v_out = torch.mean(variable*variable)
```

Now, we will be printing the results for all parameters:

```
#print the results
print(t_out)
print(v_out)
#result will be 7.5

v_out.backward() # backpropagation from v_out
# v_out = 1/4 * sum(variable*variable)
# the gradients with respect to the variable,

#Let's print the variable gradient

print(variable.grad)
'''
 0.5000 1.0000
 1.5000 2.0000
'''

print("Resultant data in the variable: "+str(variable)) # this is data in
variable

"""
Variable containing:
 1 2
 3 4
We will consider the variable as a FloatTensor
[torch.FloatTensor of size 2x2]
"""

print(variable.data) # this is data in tensor format
"""
 1 2
```

```
   3 4
We will consider the variable as FloatTensor
[torch.FloatTensor of size 2x2]
"""

#we will print the result in the numpy format
print(variable.data.numpy())
"""
[[ 1. 2.]
 [ 3. 4.]]
"""
```

Here is the output of the preceding code block:

```
tensor([[1., 2., 3.],
        [4., 5., 6.]])
tensor([[1., 2., 3.],
        [4., 5., 6.]], requires_grad=True)
tensor(15.1667)
tensor(15.1667, grad_fn=<MeanBackward1>)
tensor([[0.3333, 0.6667, 1.0000],
        [1.3333, 1.6667, 2.0000]])
Data in the variabletensor([[1., 2., 3.],
        [4., 5., 6.]], requires_grad=True)
tensor([[1., 2., 3.],
        [4., 5., 6.]])
[[1. 2. 3.]
 [4. 5. 6.]]
```

Now, let's try plotting data on a graph using `matplotlib`.

Plotting values on a graph

Let's work on one simple program to plot values on a graph. To do this, use the following code:

```
#This line is necessary to print the output inside jupyter notebook
%matplotlib inline

import torch
import matplotlib.pyplot as plt
import torch.nn.functional as F
from torch.autograd import Variable


# dummy data for the example
#lets declare linspace
```

```
x = torch.linspace(-5, 5, 200) # x data (tensor), shape=(100, 1)
x = Variable(x)
#call numpy array to plot the results
x_np = x.data.numpy()
```

Following code block lists down a few of the activation methods:

```
#RelU function
y_relu = torch.relu(x).data.numpy()
#sigmoid method
y_sigmoid = torch.sigmoid(x).data.numpy()
#tanh method
y_tanh = torch.tanh(x).data.numpy()
#softplus method
y_softplus = F.softplus(x).data.numpy() # there's no softplus in torch
# y_softmax = torch.softmax(x, dim=0).data.numpy() softmax is an activation
function and it deals with probability
```

Using `matplotlib` to activate the functions:

```
#we will plot the activation function with matplotlib
plt.figure(1, figsize=(8, 6))
plt.subplot(221)
plt.plot(x_np, y_relu, c='red', label='relu')
plt.ylim((-1, 5))
plt.legend(loc='best')

#sigmoid activation function
plt.subplot(222)
plt.plot(x_np, y_sigmoid, c='red', label='sigmoid')
plt.ylim((-0.2, 1.2))
plt.legend(loc='best')

#tanh activation function
plt.subplot(223)
plt.plot(x_np, y_tanh, c='red', label='tanh')
plt.ylim((-1.2, 1.2))
plt.legend(loc='best')

#softplus activation function
plt.subplot(224)
plt.plot(x_np, y_softplus, c='red', label='softplus')
plt.ylim((-0.2, 6))
plt.legend(loc='best')

#call the show method to draw the graph on screen
plt.show()
```

Let's plot the values on the graph, as shown in the following screenshot:

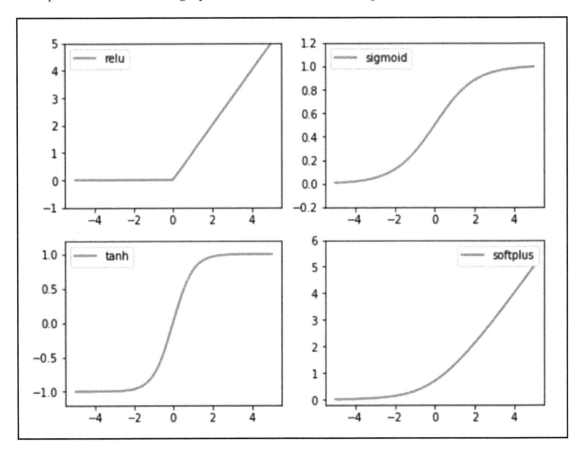

Note that the first line in the preceding code is required to draw the graph inside Jupyter Notebook. If you are running the Python file directly from the Terminal, you can omit the first line of the code.

Building our own model network

In this section, we will work on building our own network using PyTorch with a step-by-step example.

Let's begin by looking at linear regression as a starting point.

Linear regression

Linear regression is probably the first method that anyone will learn in terms of machine learning. The objective of linear regression is to find a relationship between one or more features (independent variables) and a continuous target variable (the dependent variable), which can be seen in the following code.

Import all the necessary libraries and declare all the necessary variables:

```
%matplotlib inline

#Import all the necessary libraries
import torch
import torch.nn.functional as F
import matplotlib.pyplot as plt


#we will define data points for both x-axis and y-axis
# x data (tensor), shape=(100, 1)
x = torch.unsqueeze(torch.linspace(-1, 1, 100), dim=1)
# noisy y data (tensor), shape=(100, 1)
y = x.pow(2) + 0.2*torch.rand(x.size())

# torch can only train on Variable, so convert them to Variable
# x, y = Variable(x), Variable(y)

# plt.scatter(x.data.numpy(), y.data.numpy())
# plt.show()
```

We will define the linear regression class and run a simple nn to explain regression:

```
class Net(torch.nn.Module):
 def __init__(self, n_feature, n_hidden, n_output):
  super(Net, self).__init__()
  self.hidden = torch.nn.Linear(n_feature, n_hidden) # hidden layer
  self.predict = torch.nn.Linear(n_hidden, n_output) # output layer

 def forward(self, x):
  x = F.relu(self.hidden(x)) # activation function for hidden layer
```

```
    x = self.predict(x) # linear output
    return x

net = Net(n_feature=1, n_hidden=10, n_output=1) # define the network
print(net) # net architecture

optimizer = torch.optim.SGD(net.parameters(), lr=0.2)
loss_func = torch.nn.MSELoss() # this is for regression mean squared loss

plt.ion() # something about plotting

for t in range(200):
 prediction = net(x) # input x and predict based on x
 loss = loss_func(prediction, y) # must be (1. nn output, 2. target)
 optimizer.zero_grad() # clear gradients for next train
 loss.backward() # backpropagation, compute gradients
 optimizer.step() # apply gradients
 if t % 50 == 0:
```

Now we will see how to plot the graphs and display the process of learning:

```
    plt.cla()
    plt.scatter(x.data.numpy(), y.data.numpy())
    plt.plot(x.data.numpy(), prediction.data.numpy(), 'r-', lw=5)
    plt.text(0.5, 0, 'Loss=%.4f' % loss.data.numpy(), fontdict={'size':
20, 'color': 'black'})
    plt.pause(0.1)

plt.ioff()
plt.show()
```

Let's plot the output of this code on the graph, as follows:

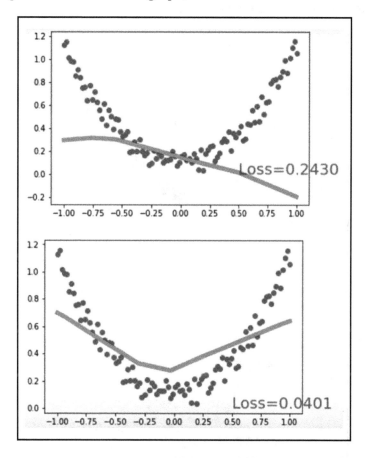

The final plot looks as follows, with the loss (meaning the deviation between the predicted output and the actual output) equaling 0.01:

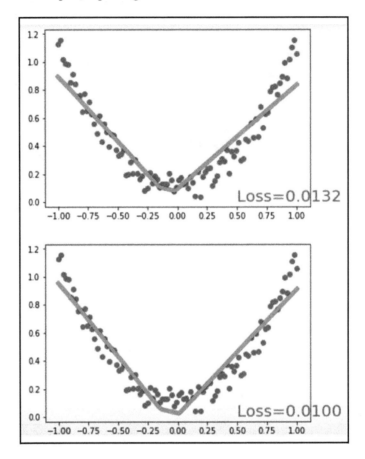

Now, we will start working toward deeper use cases using PyTorch.

Classification

A classification problem runs a neural network model to classify the inputs. For example, it classifies images of clothing into trousers, tops, and shirts. When we provide more inputs to the classification model, it will predict the value of the outcomes.

A simple example would be filtering an email as *spam* or *not spam*. Classification either predicts categorical class labels based on the training set or the values (class labels) when classifying attributes that are used in classifying new data. There are many classification models, such as Naive Bayes, random forests, decision tress, and logistic regression.

Here, we will work on a simple classification problem. To do this, use the following this code:

```
%matplotlib inline

import torch
import torch.nn.functional as F
import matplotlib.pyplot as plt

# torch.manual_seed(1) # reproducible

# make fake data
n_data = torch.ones(100, 2)
x0 = torch.normal(2*n_data, 1) # class0 x data (tensor), shape=(100, 2)
y0 = torch.zeros(100) # class0 y data (tensor), shape=(100, 1)
x1 = torch.normal(-2*n_data, 1) # class1 x data (tensor), shape=(100, 2)
y1 = torch.ones(100) # class1 y data (tensor), shape=(100, 1)
x = torch.cat((x0, x1), 0).type(torch.FloatTensor) # shape (200, 2)
FloatTensor = 32-bit floating
y = torch.cat((y0, y1), ).type(torch.LongTensor) # shape (200,) LongTensor
= 64-bit integer

class Net(torch.nn.Module):
 def __init__(self, n_feature, n_hidden, n_output):
 super(Net, self).__init__()
 self.hidden = torch.nn.Linear(n_feature, n_hidden) # hidden layer
 self.out = torch.nn.Linear(n_hidden, n_output) # output layer

def forward(self, x):
 x = F.relu(self.hidden(x)) # activation function for hidden layer
 x = self.out(x)
 return x

net = Net(n_feature=2, n_hidden=10, n_output=2) # define the network
print(net) # net architecture

optimizer = torch.optim.SGD(net.parameters(), lr=0.02)
loss_func = torch.nn.CrossEntropyLoss() # the target label is NOT an one-
hotted

plt.ion() # something about plotting
```

```
for t in range(100):
 out = net(x) # input x and predict based on x
 loss = loss_func(out, y) # must be (1. nn output, 2. target), the target
label is NOT one-hotted

optimizer.zero_grad() # clear gradients for next train
 loss.backward() # backpropagation, compute gradients
 optimizer.step() # apply gradients

if t % 10 == 0:
```

Now, let's plot the graphs and display the learning processes:

```
plt.cla()
prediction = torch.max(out, 1)[1]
pred_y = prediction.data.numpy()
target_y = y.data.numpy()
plt.scatter(x.data.numpy()[:, 0], x.data.numpy()[:, 1], c=pred_y, s=100,
lw=0, cmap='RdYlGn')
 accuracy = float((pred_y == target_y).astype(int).sum()) /
float(target_y.size)
 plt.text(1.5, -4, 'Accuracy=%.2f' % accuracy, fontdict={'size': 20,
'color': 'red'})
 plt.pause(0.1)

plt.ioff()
plt.show()
```

The output of the preceding code is as follows:

```
Net(
   (hidden): Linear(in_features=2, out_features=10, bias=True)
   (out): Linear(in_features=10, out_features=2, bias=True)
)
```

We will pick only a few plots from the output, as shown in the following screenshot:

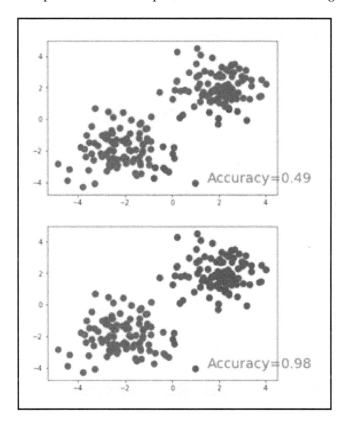

You can see that the accuracy levels have increased with the increased number of steps in the iteration:

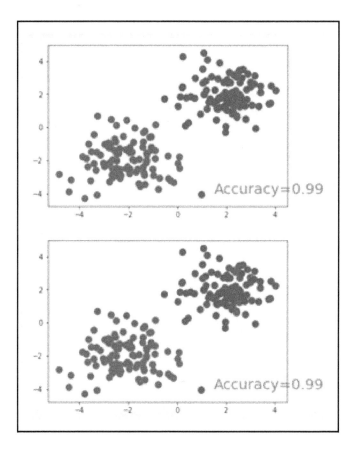

We can reach an accuracy level of 1.00 in the final step of our execution:

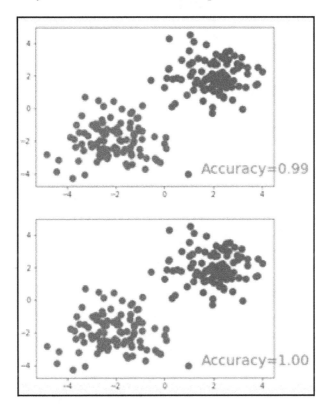

Simple neural networks with torch

Neural networks are necessary when a heuristic approach is required to solve a problem. Let's explore a basic neural network using the following example:

```python
import torch
import torch.nn.functional as F

# replace following class code with an easy sequential network
class Net(torch.nn.Module):
 def __init__(self, n_feature, n_hidden, n_output):
 super(Net, self).__init__()
 self.hidden = torch.nn.Linear(n_feature, n_hidden) # hidden layer
 self.predict = torch.nn.Linear(n_hidden, n_output) # output layer

 def forward(self, x):
 x = F.relu(self.hidden(x)) # activation function for hidden layer
```

```
x = self.predict(x)  # linear output
return x

net1 = Net(1, 10, 1)
```

Following is the easiest and fastest way to build your network:

```
net2 = torch.nn.Sequential(
torch.nn.Linear(1, 10),
torch.nn.ReLU(),
torch.nn.Linear(10, 1)
)

print(net1)  # net1 architecture
"""
Net (
  (hidden): Linear (1 -> 10)
  (predict): Linear (10 -> 1)
)
"""

print(net2)  # net2 architecture
"""
Sequential (
  (0): Linear (1 -> 10)
  (1): ReLU ()
  (2): Linear (10 -> 1)
)
"""
```

The output of the preceding code is as follows:

```
Net(
    (hidden): Linear(in_features=1, out_features=10, bias=True)
    (predict): Linear(in_features=10, out_features=1, bias=True)
)
Sequential(
    (0): Linear(in_features=1, out_features=10, bias=True)
    (1): ReLU()
    (2): Linear(in_features=10, out_features=1, bias=True)
)

Out[1]:
'\nSequential (\n  (0): Linear (1 -> 10)\n  (1): ReLU ()\n  (2): Linear (10
-> 1)\n)\n'
```

Saving and reloading data on the network

Let's look at one example of how to save data on the network and then restore the data:

```python
%matplotlib inline

import torch
import matplotlib.pyplot as plt

# torch.manual_seed(1) # reproducible

# fake data
x = torch.unsqueeze(torch.linspace(-1, 1, 100), dim=1) # x data (tensor),
shape=(100, 1)
y = x.pow(2) + 0.2*torch.rand(x.size()) # noisy y data (tensor),
shape=(100, 1)

# The code below is deprecated in Pytorch 0.4. Now, autograd directly
supports tensors
# x, y = Variable(x, requires_grad=False), Variable(y, requires_grad=False)

def save():
 # save net1
 net1 = torch.nn.Sequential(
 torch.nn.Linear(1, 10),
 torch.nn.ReLU(),
 torch.nn.Linear(10, 1)
 )
 optimizer = torch.optim.SGD(net1.parameters(), lr=0.5)
 loss_func = torch.nn.MSELoss()

for t in range(100):
 prediction = net1(x)
 loss = loss_func(prediction, y)
 optimizer.zero_grad()
 loss.backward()
 optimizer.step()

# plot result
 plt.figure(1, figsize=(10, 3))
 plt.subplot(131)
 plt.title('Net1')
 plt.scatter(x.data.numpy(), y.data.numpy())
 plt.plot(x.data.numpy(), prediction.data.numpy(), 'r-', lw=5)
```

Two ways to save the net:

```
torch.save(net1, 'net.pkl') # save entire net
torch.save(net1.state_dict(), 'net_params.pkl') # save only the parameters

def restore_net():
 # restore entire net1 to net2
 net2 = torch.load('net.pkl')
 prediction = net2(x)

# plot result
 plt.subplot(132)
 plt.title('Net2')
 plt.scatter(x.data.numpy(), y.data.numpy())
 plt.plot(x.data.numpy(), prediction.data.numpy(), 'r-', lw=5)


def restore_params():
 # restore only the parameters in net1 to net3
 net3 = torch.nn.Sequential(
 torch.nn.Linear(1, 10),
 torch.nn.ReLU(),
 torch.nn.Linear(10, 1)
 )

# copy net1's parameters into net3
 net3.load_state_dict(torch.load('net_params.pkl'))
 prediction = net3(x)
```

Plotting the results:

```
# plot result
 plt.subplot(133)
 plt.title('Net3')
 plt.scatter(x.data.numpy(), y.data.numpy())
 plt.plot(x.data.numpy(), prediction.data.numpy(), 'r-', lw=5)
 plt.show()

# save net1
save()

# restore entire net (may slow)
restore_net()

# restore only the net parameters
restore_params()
```

The output of the code will look similar to the graphs that are shown in the following diagram:

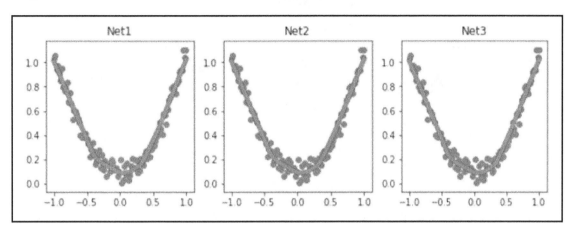

Running with batches

Torch helps you organize your data through `DataLoader`. We can use it to package the data through batch training. We can have our own data format (NumPy array, for example, or any other) loaded into Tensor, along with a wrapper.

The following is an example of a dataset where random numbers are taken into the dataset in batches and trained:

```
import torch
import torch.utils.data as Data

torch.manual_seed(1) # reproducible

BATCH_SIZE = 5

x = torch.linspace(1, 10, 10) # this is x data (torch tensor)
y = torch.linspace(10, 1, 10) # this is y data (torch tensor)

torch_dataset = Data.TensorDataset(x, y)
loader = Data.DataLoader(
  dataset=torch_dataset, # torch TensorDataset format
  batch_size=BATCH_SIZE, # mini batch size
  shuffle=True, # random shuffle for training
  num_workers=2, # subprocesses for loading data
)
```

```
def show_batch():
 for epoch in range(3): # train entire dataset 3 times
 for step, (batch_x, batch_y) in enumerate(loader): # for each training
step
 # train your data...
 print('Epoch: ', epoch, '| Step: ', step, '| batch x: ',
 batch_x.numpy(), '| batch y: ', batch_y.numpy())

if __name__ == '__main__':
 show_batch()
```

The output of the code is as follows:

```
Epoch:  0 | Step:  0 | batch x:  [ 5.  7. 10.  3.  4.] | batch y:  [6. 4.
1. 8. 7.]
Epoch:  0 | Step:  1 | batch x:  [2. 1. 8. 9. 6.] | batch y:  [ 9. 10.  3.
2.  5.]
Epoch:  1 | Step:  0 | batch x:  [ 4.  6.  7. 10.  8.] | batch y:  [7. 5.
4. 1. 3.]
Epoch:  1 | Step:  1 | batch x:  [5. 3. 2. 1. 9.] | batch y:  [ 6.  8.  9.
10.  2.]
Epoch:  2 | Step:  0 | batch x:  [ 4.  2.  5.  6. 10.] | batch y:  [7. 9.
6. 5. 1.]
Epoch:  2 | Step:  1 | batch x:  [3. 9. 1. 8. 7.] | batch y:  [ 8.  2. 10.
3.  4.]
```

Optimization algorithms

There is always doubt about which optimization algorithm should be used in our implementation of the neural network for a better output. This is done by modifying the key parameters, such as the **weights** and **bias** values.

These algorithms are used to minimize (or maximize) error ($E(x)$), which is dependent on the internal parameters. They are used for computing the target results (Y) from the set of predictors (x) that are used in the model.

Now, let's look at the different types of algorithms by using the following example:

```
%matplotlib inline

import torch
import torch.utils.data as Data
import torch.nn.functional as F
import matplotlib.pyplot as plt

# torch.manual_seed(1) # reproducible
```

```
LR = 0.01
BATCH_SIZE = 32
EPOCH = 12

# dummy dataset
x = torch.unsqueeze(torch.linspace(-1, 1, 1000), dim=1)
y = x.pow(2) + 0.1*torch.normal(torch.zeros(*x.size()))

# plot dataset
plt.scatter(x.numpy(), y.numpy())
plt.show()
```

Putting dateset into torch dataset:

```
torch_dataset = Data.TensorDataset(x, y)
loader = Data.DataLoader(dataset=torch_dataset, batch_size=BATCH_SIZE,
shuffle=True, num_workers=2,)

# default network
class Net(torch.nn.Module):
 def __init__(self):
 super(Net, self).__init__()
 self.hidden = torch.nn.Linear(1, 20) # hidden layer
 self.predict = torch.nn.Linear(20, 1) # output layer

def forward(self, x):
 x = F.relu(self.hidden(x)) # activation function for hidden layer
 x = self.predict(x) # linear output
 return x

if __name__ == '__main__':
 # different nets
 net_SGD = Net()
 net_Momentum = Net()
 net_RMSprop = Net()
 net_Adam = Net()
 nets = [net_SGD, net_Momentum, net_RMSprop, net_Adam]

# different optimizers
 opt_SGD = torch.optim.SGD(net_SGD.parameters(), lr=LR)
 opt_Momentum = torch.optim.SGD(net_Momentum.parameters(), lr=LR,
momentum=0.8)
 opt_RMSprop = torch.optim.RMSprop(net_RMSprop.parameters(), lr=LR,
alpha=0.9)
 opt_Adam = torch.optim.Adam(net_Adam.parameters(), lr=LR, betas=(0.9,
0.99))
 optimizers = [opt_SGD, opt_Momentum, opt_RMSprop, opt_Adam]
```

```
loss_func = torch.nn.MSELoss()
losses_his = [[], [], [], []] # record loss
```

Training the model for various epochs:

```
for epoch in range(EPOCH):
print('Epoch: ', epoch)
for step, (b_x, b_y) in enumerate(loader): # for each training step
for net, opt, l_his in zip(nets, optimizers, losses_his):
output = net(b_x) # get output for every net
loss = loss_func(output, b_y) # compute loss for every net
opt.zero_grad() # clear gradients for next train
loss.backward() # backpropagation, compute gradients
opt.step() # apply gradients
l_his.append(loss.data.numpy()) # loss recoder

labels = ['SGD', 'Momentum', 'RMSprop', 'Adam']
for i, l_his in enumerate(losses_his):
plt.plot(l_his, label=labels[i])
plt.legend(loc='best')
plt.xlabel('Steps')
plt.ylabel('Loss')
plt.ylim((0, 0.2))
plt.show()
```

The output of executing the preceding code block is displayed in the following plot:

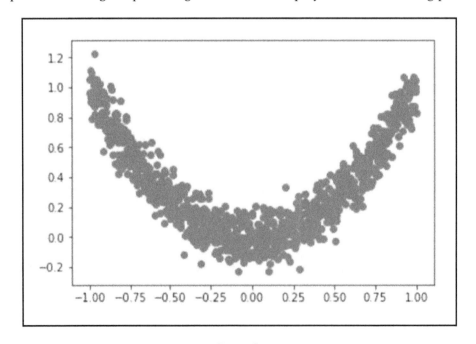

The output of the Epoch count will look like this:

```
Epoch:  0
Epoch:  1
Epoch:  2
Epoch:  3
Epoch:  4
Epoch:  5
Epoch:  6
Epoch:  7
Epoch:  8
Epoch:  9
Epoch:  10
Epoch:  11
```

We will plot all the optimizers and represent them in the graph, as follows:

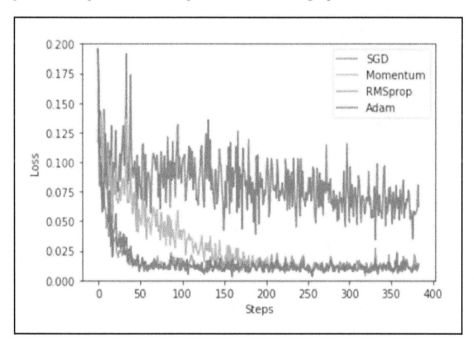

In the next section, we will look at RNNs.

Recurrent neural networks

With RNNs, unlike feedforward neural networks, we can use the internal memory to process inputs in a sequential manner. In RNN, the connection between nodes forms a directed graph along a temporal sequence. This helps in tasking the RNN with largely unsegmented and connected speech or character recognition.

The MNIST database

The **MNIST** database consists of 60,000 handwritten digits. It also consists of a test dataset that's made up of 10,000 digits. While it is a subset of the NIST dataset, all the digits in this dataset are size-normalized and have been centered on a 28 x 28 pixels-sized image. Here, every pixel contains a value of 0-255 with its grayscale value.

The MNIST dataset can be found at `http://yann.lecun.com/exdb/mnist/`.

The NIST dataset can be found a `https://www.nist.gov/srd/nist-special-database-19`.

RNN classification

Here, we will look at an example of how to build an RNN to identify handwritten numbers from the MNIST database:

```
import torch
from torch import nn
import torchvision.datasets as dsets
import torchvision.transforms as transforms
import matplotlib.pyplot as plt

# torch.manual_seed(1) # reproducible

# Hyper Parameters
EPOCH = 1 # train the training data n times, to save time, we just train 1
epoch
BATCH_SIZE = 64
TIME_STEP = 28 # rnn time step / image height
INPUT_SIZE = 28 # rnn input size / image width
LR = 0.01 # learning rate
DOWNLOAD_MNIST = True # set to True if haven't download the data

# Mnist digital dataset
train_data = dsets.MNIST(
```

```
  root='./mnist/',
  train=True, # this is training data
  transform=transforms.ToTensor(), # Converts a PIL.Image or numpy.ndarray
to
  # torch.FloatTensor of shape (C x H x W) and normalize in the range [0.0,
1.0]
  download=DOWNLOAD_MNIST, # download it if you don't have it
)
```

Plotting one example:

```
print(train_data.train_data.size()) # (60000, 28, 28)
print(train_data.train_labels.size()) # (60000)
plt.imshow(train_data.train_data[0].numpy(), cmap='gray')
plt.title('%i' % train_data.train_labels[0])
plt.show()

# Data Loader for easy mini-batch return in training
train_loader = torch.utils.data.DataLoader(dataset=train_data,
batch_size=BATCH_SIZE, shuffle=True)
```

Converting test data into Variable, pick 2000 samples to speed up testing:

```
test_data = dsets.MNIST(root='./mnist/', train=False,
transform=transforms.ToTensor())
test_x = test_data.test_data.type(torch.FloatTensor)[:2000]/255. # shape
(2000, 28, 28) value in range(0,1)
test_y = test_data.test_labels.numpy()[:2000] # covert to numpy array

class RNN(nn.Module):
 def __init__(self):
 super(RNN, self).__init__()

self.rnn = nn.LSTM( # if use nn.RNN(), it hardly learns
 input_size=INPUT_SIZE,
 hidden_size=64, # rnn hidden unit
 num_layers=1, # number of rnn layer
 batch_first=True, # input & output will has batch size as 1s dimension.
e.g. (batch, time_step, input_size)
 )

self.out = nn.Linear(64, 10)

def forward(self, x):
 # x shape (batch, time_step, input_size)
 # r_out shape (batch, time_step, output_size)
 # h_n shape (n_layers, batch, hidden_size)
 # h_c shape (n_layers, batch, hidden_size)
```

```
    r_out, (h_n, h_c) = self.rnn(x, None) # None represents zero initial
hidden state

# choose r_out at the last time step
 out = self.out(r_out[:, -1, :])
 return out

rnn = RNN()
print(rnn)

optimizer = torch.optim.Adam(rnn.parameters(), lr=LR) # optimize all cnn
parameters
loss_func = nn.CrossEntropyLoss() # the target label is not one-hotted
```

Training and testing the epochs:

```
for epoch in range(EPOCH):
 for step, (b_x, b_y) in enumerate(train_loader): # gives batch data
 b_x = b_x.view(-1, 28, 28) # reshape x to (batch, time_step, input_size)

output = rnn(b_x) # rnn output
 loss = loss_func(output, b_y) # cross entropy loss
 optimizer.zero_grad() # clear gradients for this training step
 loss.backward() # backpropagation, compute gradients
 optimizer.step() # apply gradients

if step % 50 == 0:
 test_output = rnn(test_x) # (samples, time_step, input_size)
 pred_y = torch.max(test_output, 1)[1].data.numpy()
 accuracy = float((pred_y == test_y).astype(int).sum()) /
float(test_y.size)
 print('Epoch: ', epoch, '| train loss: %.4f' % loss.data.numpy(), '| test
accuracy: %.2f' % accuracy)

# print 10 predictions from test data
test_output = rnn(test_x[:10].view(-1, 28, 28))
pred_y = torch.max(test_output, 1)[1].data.numpy()
print(pred_y, 'prediction number')
print(test_y[:10], 'real number')
```

The following files need to be downloaded and extracted to train the images:

```
Downloading http://yann.lecun.com/exdb/mnist/train-images-idx3-ubyte.gz to
./mnist/MNIST/raw/train-images-idx3-ubyte.gz
100.1%
Extracting ./mnist/MNIST/raw/train-images-idx3-ubyte.gz
Downloading http://yann.lecun.com/exdb/mnist/train-labels-idx1-ubyte.gz to
./mnist/MNIST/raw/train-labels-idx1-ubyte.gz
```

```
113.5%
Extracting ./mnist/MNIST/raw/train-labels-idx1-ubyte.gz
Downloading http://yann.lecun.com/exdb/mnist/t10k-images-idx3-ubyte.gz to
./mnist/MNIST/raw/t10k-images-idx3-ubyte.gz
100.4%
Extracting ./mnist/MNIST/raw/t10k-images-idx3-ubyte.gz
Downloading http://yann.lecun.com/exdb/mnist/t10k-labels-idx1-ubyte.gz to
./mnist/MNIST/raw/t10k-labels-idx1-ubyte.gz
180.4%
Extracting ./mnist/MNIST/raw/t10k-labels-idx1-ubyte.gz
Processing...
Done!
torch.Size([60000, 28, 28])
torch.Size([60000])
/usr/local/lib/python3.7/site-packages/torchvision/datasets/mnist.py:53:
UserWarning: train_data has been renamed data
  warnings.warn("train_data has been renamed data")
/usr/local/lib/python3.7/site-packages/torchvision/datasets/mnist.py:43:
UserWarning: train_labels has been renamed targets
  warnings.warn("train_labels has been renamed targets")
```

The output of the preceding code is as follows:

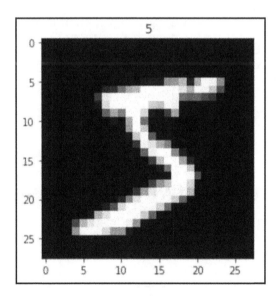

Let's take the processing further with this code:

```
/usr/local/lib/python3.7/site-packages/torchvision/datasets/mnist.py:58:
UserWarning: test_data has been renamed data
```

```
  warnings.warn("test_data has been renamed data")
/usr/local/lib/python3.7/site-packages/torchvision/datasets/mnist.py:48:
UserWarning: test_labels has been renamed targets
  warnings.warn("test_labels has been renamed targets")

RNN(
  (rnn): LSTM(28, 64, batch_first=True)
  (out): Linear(in_features=64, out_features=10, bias=True)
)
```

The output of epochs is as follows:

```
Epoch:  0 | train loss: 2.3156 | test accuracy: 0.12
Epoch:  0 | train loss: 1.1875 | test accuracy: 0.57
Epoch:  0 | train loss: 0.7739 | test accuracy: 0.68
Epoch:  0 | train loss: 0.8689 | test accuracy: 0.73
Epoch:  0 | train loss: 0.5322 | test accuracy: 0.83
Epoch:  0 | train loss: 0.3657 | test accuracy: 0.83
Epoch:  0 | train loss: 0.2960 | test accuracy: 0.88
Epoch:  0 | train loss: 0.3869 | test accuracy: 0.90
Epoch:  0 | train loss: 0.1694 | test accuracy: 0.92
Epoch:  0 | train loss: 0.0869 | test accuracy: 0.93
Epoch:  0 | train loss: 0.2825 | test accuracy: 0.91
Epoch:  0 | train loss: 0.2392 | test accuracy: 0.94
Epoch:  0 | train loss: 0.0994 | test accuracy: 0.91
Epoch:  0 | train loss: 0.3731 | test accuracy: 0.94
Epoch:  0 | train loss: 0.0959 | test accuracy: 0.94
Epoch:  0 | train loss: 0.1991 | test accuracy: 0.95
Epoch:  0 | train loss: 0.0711 | test accuracy: 0.94
Epoch:  0 | train loss: 0.2882 | test accuracy: 0.96
Epoch:  0 | train loss: 0.4420 | test accuracy: 0.95
[7 2 1 0 4 1 4 9 5 9] prediction number
[7 2 1 0 4 1 4 9 5 9] real number
```

RNN cyclic neural network – regression

Now, we will deal with a regression problem under RNN. The cyclic neural network provides memory to the neural network. For the serial data, the cyclic neural network can achieve better results. We will use RNN here in this example to predict time series data.

To find out more about circular neural networks, go to `https://iopscience.iop.org/article/10.1209/0295-5075/18/3/003/meta`.

The following code is for the logistic regression:

```
%matplotlib inline

import torch
from torch import nn
import numpy as np
import matplotlib.pyplot as plt

# torch.manual_seed(1) # reproducible

# Hyper Parameters
TIME_STEP = 10 # rnn time step
INPUT_SIZE = 1 # rnn input size
LR = 0.02 # learning rate

# show data
steps = np.linspace(0, np.pi*2, 100, dtype=np.float32) # float32 for
converting torch FloatTensor
x_np = np.sin(steps)
y_np = np.cos(steps)
plt.plot(steps, y_np, 'r-', label='target (cos)')
plt.plot(steps, x_np, 'b-', label='input (sin)')
plt.legend(loc='best')
plt.show()
```

The RNN class is defined in the following code. We will use r_out in a linear way to calculated the predicted output. We can also use a for loop to calculate the predicted output with torch.stack:

```
class RNN(nn.Module):
 def __init__(self):
     super(RNN, self).__init__()

 self.rnn = nn.RNN(
 input_size=INPUT_SIZE,
 hidden_size=32, # rnn hidden unit
 num_layers=1, # number of rnn layer
 batch_first=True, # input & output will have batch size as 1s dimension.
e.g. (batch, time_step, input_size)
 )
 self.out = nn.Linear(32, 1)

 def forward(self, x, h_state):
     # x (batch, time_step, input_size)
     # h_state (n_layers, batch, hidden_size)
     # r_out (batch, time_step, hidden_size)
     r_out, h_state = self.rnn(x, h_state)
```

```
      outs = [] # save all predictions
      for time_step in range(r_out.size(1)):
  outs.append(self.out(r_out[:, time_step, :]))
      return torch.stack(outs, dim=1), h_state
//instantiate RNN
rnn = RNN()
print(rnn)
```

The output is as follows:

```
"""
RNN (
  (rnn): RNN(1, 32, batch_first=True)
  (out): Linear (32 -> 1)
)
"""
```

We need to optimize RNN parameters now, as shown in the following code, before running the `for` loop to give the prediction:

```
optimizer = torch.optim.Adam(rnn.parameters(), lr=LR)
loss_func = nn.MSELoss()
h_state = None
plt.figure(1, figsize=(12, 5))
plt.ion()
```

The following block of code will look like a motion picture effect when it's run, which we can't represent here in this book. We have added a few screenshots to help you visualize this. We are using x as an input `sin` value and y as an output fitting `cos` value. Because a relationship exists between the two curves, we will use `sin` to predict `cos`:

```
for step in range(100):
  start, end = step * np.pi, (step+1)*np.pi # time range
  # use sin predicts cos
  steps = np.linspace(start, end, TIME_STEP, dtype=np.float32,
endpoint=False) # float32 for converting torch FloatTensor
  x_np = np.sin(steps)
  y_np = np.cos(steps)

  x = torch.from_numpy(x_np[np.newaxis, :, np.newaxis]) # shape (batch,
time_step, input_size)
  y = torch.from_numpy(y_np[np.newaxis, :, np.newaxis])

  prediction, h_state = rnn(x, h_state) # rnn output

  h_state = h_state.data # repack the hidden state, break the connection from
last iteration
```

```
loss = loss_func(prediction, y) # calculate loss
optimizer.zero_grad() # clear gradients for this training step
loss.backward() # backpropagation, compute gradients
optimizer.step() # apply gradients
```

Plotting the results:

```
plt.plot(steps, y_np.flatten(), 'r-')
plt.plot(steps, prediction.data.numpy().flatten(), 'b-')
plt.draw(); plt.pause(0.05)

plt.ioff()
plt.show()
```

The output of the preceding code is as follows:

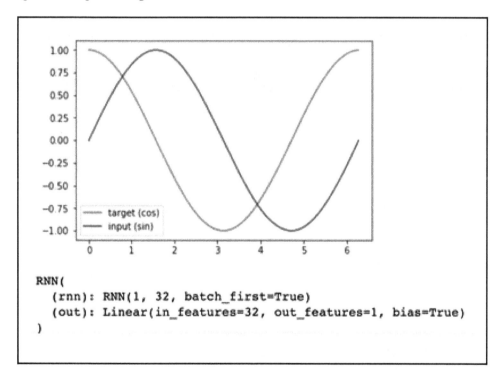

The following is a plot graph that will be generated after iteration 10:

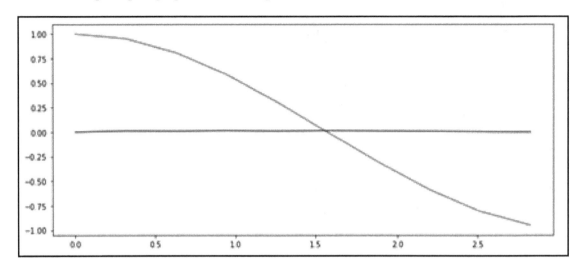

The following is a plot graph that will be generated after iteration 25:

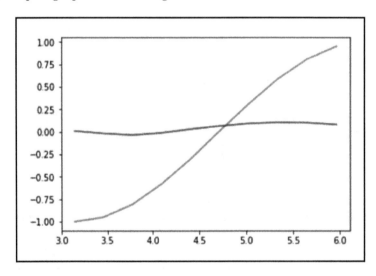

We are not showing all 100 iteration output images here, but we will skip to the final output, iteration 100, as shown in the following screenshot:

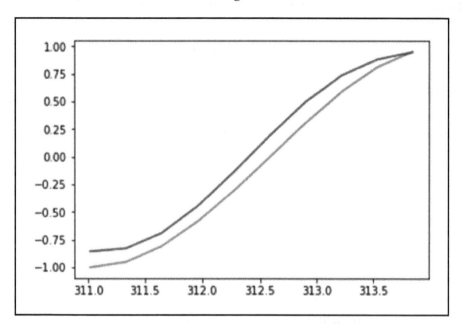

In the next section, we will look at NLP.

Natural language processing

Now is the time to experiment with a few NLP techniques with the help of PyTorch. This will be more useful for those of you who haven't written code in any deep learning framework before, but who may have better understanding of NLP core problems and algorithms.

In this chapter, we will look into simple examples with small dimensions, so that we can see how the weight of the layers changes as the network is training. You can try out your own model once you understand the network and how it works.

Before working on any NLP-based problems, we need to understand the basic building blocks on deep learning, including affine maps, non-linearities, and objective functions.

Affine maps

Affine maps are one of the basic building components of deep learning, and are represented as follows:

$$f(x) = Ax + b$$

In this case, the matrix is represented by A and the vectors are represented by x and b. A and b are the parameters that need to be learned, while b is the bias.

A simple example to explain this is as follows:

```
import torch
import torch.nn as nn
import torch.nn.functional as F
import torch.optim as optim

torch.manual_seed(1)
lin = nn.Linear(6, 3) # maps from R^6 to R^3, parameters A, b
# data is 2x5. A maps from 6 to 3... can we map "data" under A?
data = torch.randn(2, 6)
print(lin(data)
```

After this, run the program with the following command:

```
$ python3 torch_nlp.py
```

The output will be as follows:

```
tensor([[ 1.1105, -0.1102, -0.3235],
        [ 0.4800,  0.1633, -0.2515]], grad_fn=<AddmmBackward>)
```

Non-linearities

First, we need to identify why we need non-linearities. Consider, we have two affine maps: `f(x)=Ax+b` and `g(x)=Cx+d`. `f(g(x))` is shown in the following equation:

$$f(g(x)) = A(Cx + d) + b = ACx + (Ad + b)$$

Here, we can see that when affine maps are composed together, the resultant is an affine map, where $Ad+b$ is a vector and AC is a matrix.

We can identify neural networks as long chains of affine compositions. Previously, it was possible that non-linearities were introduced in-between the affine layers. But thankfully, it isn't the case any longer, and hence that helps in building more powerful and efficient models.

While working with the most common functions such as tanh (x), σ(x) and ReLU (x), we see that there are a few core non-linearities, as shown in the following code block:

```
#let's see more about non-linearities
#Most of the non-linearities in PyTorch are present in torch.functional
which we import as F)
# Please make a note that unlike affine maps, there are mostly no
parameters in non-linearites
# That is, they don't have weights that are updated during training.
#This means that during training the weights are not updated.
data = torch.randn(2, 2)
print(data)
print(F.relu(data))
```

The output of the preceding code is as follows:

```
tensor([[ 0.5848, 0.2149],
  [-0.4090, -0.1663]])
tensor([[0.5848, 0.2149],
  [0.0000, 0.0000]])
```

Objective functions

The objective function (also called the loss function or cost function) will help your network to minimize. It works by selecting a training instance, running it through your neural network, then computing the loss of the output.

The derivative of the loss function is updated for finding the parameters of the model. Like, if your model predicts an answer confidently, and the answer turns out to be wrong, the the computed loss will be high. If the predicted answer is correct, then the loss is low.

How is the network minimized?

1. First, the function will select a training instance
2. Then, it is passed through our neural network to get the output
3. Finally, the loss of the output is calculated

In our training examples we need to minimize the loss function to minimize the probability of wrong results with the actual dataset.

Building network components in PyTorch

Before shifting our focus to NLP, in this section we will use non-linearities and affine maps to build a network in PyTorch. In this example, we will learn to compute a loss function using the built in negative log likelihood in PyTorch and using backpropagation for updating the parmeters.

Please note that all the components of the network need to be inherited from `nn.Module` and also override the `forward()` method. Considering boilerplate, these are the details we should remember. The network components are provided functionality when we inherit those components from `nn.Module`

Now, as mentioned previously, we will look at an example, in which the network takes a scattered bag-of-words (BoW) representation and and the output is a probability distribution into two labels, that is, English and Spanish. Also, this model is an example of logistic regression.

BoW classifier using logistic regression

Probabilities will be logged onto our two labels English and Spanish on which our generated model will map a sparse BoW representation. In the vocabulary, we will assign each word as an index. Let's say for example, we have two words in our vocabulary, that is hello and world, which have indices as zero and one, respectively. For example, for the sentence *hello hello hello hello hello,* the BoW vector is *[5,0]*. Similarly the BoW vector for *hello world world hello world* is *[2,3]*, and so on.

Generally, it is *[Count(hello),Count(world)]*.

Let us denote is BOW vector as *x*.

The network output is as follows:

$$logSoftmax(Ax + b)$$

Next, we need to pass the input through an affine map and then use log softmax:

```
data = [("El que lee mucho y anda mucho, ve mucho y sabe mucho".split(),
"SPANISH"),
  ("The one who reads a lot and walks a lot, sees a lot and knows a
```

```
lot.".split(), "ENGLISH"),
 ("Nunca es tarde si la dicha es buena".split(), "SPANISH"),
 ("It is never late if the joy is good".split(), "ENGLISH")]

test_data = [("Que cada palo aguante su vela".split(), "SPANISH"),
 ("May every mast hold its own sail".split(), "ENGLISH")]

#each word in the vocabulary is mapped to an unique integer using
word_to_ix, and that will be considered as that word's index in BOW

word_to_ix = {}
for sent, _ in data + test_data:
 for word in sent:
 if word not in word_to_ix:
 word_to_ix[word] = len(word_to_ix)
print(word_to_ix)

VOCAB_SIZE = len(word_to_ix)
NUM_LABELS = 2

class BoWClassifier(nn.Module): # inheriting from nn.Module!

def __init__(self, num_labels, vocab_size):

#This calls the init function of nn.Module. The syntax might confuse you,
but don't be confused. Remember to do it in nn.module

 super(BoWClassifier, self).__init__()
```

Next, we will define the parameters that are needed. Here, those parameters are A and B, and the following code block explains the further implementations are required:

```
# let's look at the prarmeters required for affine mapping
# nn.Linear() is defined using Torch that gives us the affine maps.
#We need to ensure that we understand why the input dimension is vocab_size
# num_labels is the output
self.linear = nn.Linear(vocab_size, num_labels)

# Important thing to remember: parameters are not present in the non-
linearity log softmax. So, let's now think about that.

def forward(self, bow_vec):
 #first, the input is passed through the linear layer
 #then it is passed through log_softmax
 #torch.nn.functional contains other non-linearities and many other
 fuctions

 return F.log_softmax(self.linear(bow_vec), dim=1)
```

```
def make_bow_vector(sentence, word_to_ix):
 vec = torch.zeros(len(word_to_ix))
 for word in sentence:
 vec[word_to_ix[word]] += 1
 return vec.view(1, -1)

def make_target(label, label_to_ix):
 return torch.LongTensor([label_to_ix[label]])

model = BoWClassifier(NUM_LABELS, VOCAB_SIZE)
```

Now, the model knows its own parameters. The first output is A, while the second is B, as follows:

```
#A component is assigned to a class variable in the __init__ function
# of a module, which was done with the line
# self.linear = nn.Linear(...)

# Then from the PyTorch devs, knowledge of the nn.linear's parameters #is
stored by the module (here-BoW Classifier)

for param in model.parameters():
 print(param)


#Pass a BoW vector for running the model
# the code is wrapped since we don't need to train it
torch.no_grad()
with torch.no_grad():
 sample = data[0]
 bow_vector = make_bow_vector(sample[0], word_to_ix)
 log_probs = model(bow_vector)
 print(log_probs)
```

The output of the preceding code is as follows:

```
{'El': 0, 'que': 1, 'lee': 2, 'mucho': 3, 'y': 4, 'anda': 5, 'mucho,': 6,
've': 7, 'sabe': 8, 'The': 9, 'one': 10, 'who': 11, 'reads': 12, 'a': 13,
'lot': 14, 'and': 15, 'walks': 16, 'lot,': 17, 'sees': 18, 'knows': 19,
'lot.': 20, 'Nunca': 21, 'es': 22, 'tarde': 23, 'si': 24, 'la': 25,
'dicha': 26, 'buena': 27, 'It': 28, 'is': 29, 'never': 30, 'late': 31,
'if': 32, 'the': 33, 'joy': 34, 'good': 35, 'Que': 36, 'cada': 37, 'palo':
38, 'aguante': 39, 'su': 40, 'vela': 41, 'May': 42, 'every': 43, 'mast':
44, 'hold': 45, 'its': 46, 'own': 47, 'sail': 48}
Parameter containing:
tensor([[-0.0347, 0.1423, 0.1145, -0.0067, -0.0954, 0.0870, 0.0443,
-0.0923,
 0.0928, 0.0867, 0.1267, -0.0801, -0.0235, -0.0028, 0.0209, -0.1084,
```

```
 -0.1014,  0.0777, -0.0335,  0.0698,  0.0081,  0.0469,  0.0314,  0.0519,
 0.0708, -0.1323,  0.0719, -0.1004, -0.1078,  0.0087, -0.0243,  0.0839,
-0.0827, -0.1270,  0.1040, -0.0212,  0.0804,  0.0459, -0.1071,  0.0287,
 0.0343, -0.0957, -0.0678,  0.0487,  0.0256, -0.0608, -0.0432,  0.1308,
-0.0264],
[ 0.0805,  0.0619, -0.0923, -0.1215,  0.1371,  0.0075,  0.0979,  0.0296,
 0.0459,  0.1067,  0.1355, -0.0948,  0.0179,  0.1066,  0.1035,  0.0887,
-0.1034, -0.1029, -0.0864,  0.0179,  0.1424, -0.0902,  0.0761, -0.0791,
-0.1343, -0.0304,  0.0823,  0.1326, -0.0887,  0.0310,  0.1233,  0.0947,
 0.0890,  0.1015,  0.0904,  0.0369, -0.0977, -0.1200, -0.0655, -0.0166,
-0.0876,  0.0523,  0.0442, -0.0323,  0.0549,  0.0462,  0.0872,  0.0962,
-0.0484]], requires_grad=True)
Parameter containing:
tensor([ 0.1396, -0.0165], requires_grad=True)
tensor([[-0.6171, -0.7755]])
```

We got the tensor output values. But, as we can see from the preceding code, these values aren't in correspondence to the log probability whether which is English and which corresponds to word Spanish. We need to train the model, and for that it's important to define these values to the log probabilities.

```
label_to_ix = {"SPANISH": 0, "ENGLISH": 1}
```

Let's start training our model then. We start with passing instances through the model to the those log probabilities. Then, the loss function is computed, and once the loss function is computer we calculate the gradient of this loss function. Finally, the parameters are updated with a gradient step. The nn package in PyTorch provides the loss functions. We want nn.NLLLoss() as the negative log likelihood loss. Optimization functions are also defined is torch.optim.

Here, we will just use **Stochastic Gradient Descent (SGD)**:

```
# Pass the BoW vector for running the model
# the code is wrapped since we don't need to train it
torch.no_grad()

with torch.no_grad():
 sample = data[0]
 bow_vector = make_bow_vector(sample[0], word_to_ix)
 log_probs = model(bow_vector)
 print(log_probs)

# We will run this on data that can be tested temporarily, before training,
just to check the before and after difference using touch.no_grad():

with torch.no_grad():
 for instance, label in test_data:
```

```
 bow_vec = make_bow_vector(instance, word_to_ix)
 log_probs = model(bow_vec)
 print(log_probs)


#The matrix column corresponding to "creo" is printed
print(next(model.parameters())[:, word_to_ix["mucho"]])

loss_function = nn.NLLLoss()
optimizer = optim.SGD(model.parameters(), lr=0.1)
```

We don't want to pass the training data again and again for no reason. Real datasets have multiple instances and not just 2. It is reasonable to train the model for epochs between 5 to 30.

The following code shows the range for our example:

```
for epoch in range(100):
 for instance, label in data:
 # Firstly, remember that gradients are accumulated by PyTorch
 # It's important that we clear those gradients before each instance
 model.zero_grad()


#The next step is to prepare our BOW vector and the target should be
#wrapped in also we must wrap the target in a tensor in the form of an
#integer
 # For example, as considered above, if the target word is SPANISH, #then,
the integer wrapped should be 0
#The loss function is already trained to understand that when the 0th
element among the log probabilities is the one that is in accordance to
SPANISH label

 bow_vec = make_bow_vector(instance, word_to_ix)
 target = make_target(label, label_to_ix)

# Next step is to run the forward pass
 log_probs = model(bow_vec)
```

Here, we will compute the various factors such as loss, gradient, and updating the parameters by calling the function optimizer.step():

```
 loss = loss_function(log_probs, target)
 loss.backward()
 optimizer.step()

with torch.no_grad():
 for instance, label in test_data:
 bow_vec = make_bow_vector(instance, word_to_ix)
```

```
log_probs = model(bow_vec)
print(log_probs)

# After computing and the results, we see that the index that corresponds
to Spanish has gone up, and for English is has gone down!
print(next(model.parameters())[:, word_to_ix["mucho"]])
```

The output is as follows:

```
tensor([[-0.7653, -0.6258]])
tensor([[-1.0456, -0.4331]])
tensor([-0.0071, -0.0462], grad_fn=<SelectBackward>)
tensor([[-0.1546, -1.9433]])
tensor([[-0.9623, -0.4813]])
tensor([ 0.4421, -0.4954], grad_fn=<SelectBackward>)
```

Summary

Now, we have the basic understanding of performing text-based processing using PyTorch. We also have a better understanding on how RNN works and how can we approach NLP-related problems using PyTorch.

In the upcoming chapters, we will build applications using what we have learned about neural networks and NLP. Happy coding!

TensorFlow on Mobile with Speech-to-Text with the WaveNet Model

7

In this chapter, we are going to learn how to convert audio to text using the WaveNet model. We will then build a model that will take audio and convert it into text using an Android application.

 This chapter is based on the *WaveNet: A Generative Model for Raw Audio* paper, by Aaron van den Oord, Sander Dieleman, Heiga Zen, Karen Simonyan, Oriol Vinyals, Alex Graves, Nal Kalchbrenner, Andrew Senior, and Koray Kavukcuoglu. You can find this paper at `https://arxiv.org/abs/1609.03499`.

In this chapter, we will cover the following topics:

- WaveNet and how it works
- The WaveNet architecture
- Building a model using WaveNet
- Preprocessing datasets
- Training the WaveNet network
- Transforming a speech WAV file into English text
- Building an Android application

Let's dig deeper into what Wavenet actually is.

WaveNet

WaveNet is a deep generative network that is used to generate raw audio waveforms. Sounds waves are generated by WaveNet to mimic the human voice. This generated sound is more natural than any of the currently existing text-to-speech systems, reducing the gap between system and human performance by 50%.

With a single WaveNet, we can differentiate between multiple speakers with equal fidelity. We can also switch between individual speakers based on their identity. This model is autoregressive and probabilistic, and it can be trained efficiently on thousands of audio samples per second. A single WaveNet can capture the characteristics of many different speakers with equal fidelity, and can switch between them by conditioning the speaker identity.

As shown in the movie *Her*, the long-standing dream of human-computer interaction is to allow people to talk to machines. The computer's ability to understand voices has increased tremendously over the past few years as a result of deep neural networks (for example, Google Assistant, Siri, Alexa, and Cortana). On the other hand, to generate speech with computers, a process referred to as speech synthesis or text to speech is followed. In the text-to-speech method, a large database of short sound fragments are recorded by a single speaker and then combined to form the required utterances. This process is very difficult because we can't change the speaker.

This difficulty has led to a great need for other methods of generating speech, where all the information that is needed for generating the data is stored in the parameters of the model. Additionally, using the inputs that are given to the model, we can control the contents and various attributes of speech. When speech is generated by adding sound fragments together, attribution graphs are generated.

The following is the attribution graph of speech that is generated in **1 second**:

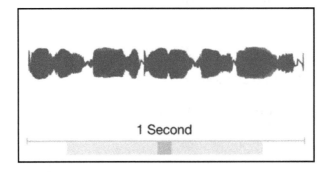

1 Second

The following is the attribution graph of speech that is generated in **100 milliseconds**:

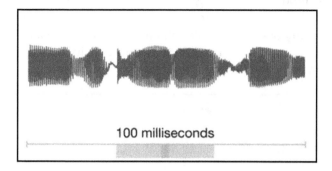

The following is the attribution graph of speech that is generated in **10 milliseconds**:

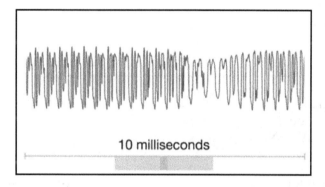

The following is the attribution graph of speech that is generated in **1 millisecond**:

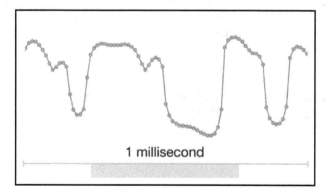

The **Pixel Recurrent Neural Network (PixelRNN)** and **Pixel Convolutional Neural Network (PixelCNN)** models from Google ensure that it's possible to generate images that include complex formations – not by generating one pixel at a time, but by an entire color channel altogether. At any one time, a color channel will need at least a thousand predictions per image. This way, we can alter a two-dimensional PixelNet into a one-dimensional WaveNet; this idea is shown in the following diagram:

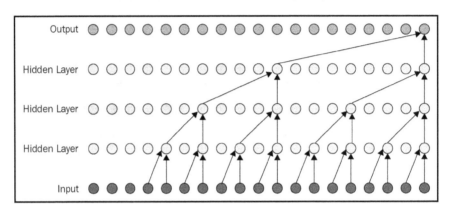

The preceding diagram displays the structure of a WaveNet model. WaveNet is a full CNN, in which the convolutional layers include a variety of dilation factors. These factors help the receptive field of WaveNet to grow exponentially with depth, and it also helps to cover thousands of time steps.

During training, the human speaker records the input sequences to create waveforms. Once the training is complete, we generate synthetic utterances by sampling the network. A value is taken from the probability distribution which is computed by the network at each step of sampling. The value that's received is fed as the input for the next step, and then a new prediction is made. Building these samples at each step is expensive; however, it's necessary to generate complex and realistic-sounding audio.

 More information about PixelRNN can be found at https://arxiv.org/pdf/1601.06759.pdf, while information about *Conditional Image Generation with PixelCNN Decoders* can be found at https://arxiv.org/pdf/1606.05328.pdf.

Architecture

The architecture of WaveNet neural networks shows amazing outputs by generating audio and text-to-speech translations, since it directly produces a raw audio waveform.

When the previous samples and additional parameters are given as the input, the network produces the next sample in the form of an audio waveform using conditional probability.

The waveform that is given as the input is quantized to a fixed range of integers. This happens after the audio is preprocessed. The tensors are produced by one-hot encoding these integer amplitudes. Hence, the dimensions of the channel are reduced by the convolutional layer that only accesses the current and previous inputs.

The following diagram displays the WaveNet architecture:

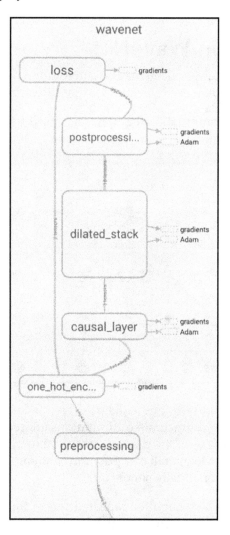

A stack of causal dilated layers is used to build the network core. Each layer is a dilated convolution with holes, and it accesses only the past and current audio samples.

Then, the outputs that are received from all the layers are combined and, using an array of dense postprocessing layers, they are fed to the original channels. Later, the softmax function converts the output into a categorical distribution.

The loss function is calculated as the cross entropy between the output for each time step and the input at the next time step.

Network layers in WaveNet

Here, we will focus on generating dilated causal convolution network layers with the filter size of two. Note that these ideas are relevant to larger filter sizes.

During this generation, the computational graph that's used to compute a single output value can be seen as a binary tree:

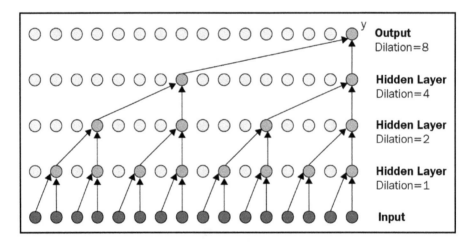

The **Input** nodes on the bottom layer of the diagram are the leaves of the tree, while the **Output** layer is the root. The intermediate computations are represented by the nodes above the **Input** layer. The edges of the graph correspond to multiple matrices. Since the computation is a binary tree, the overall computation time for the graph is $O(2^L)$. When L is large, the computation exponentially shoots up.

However, since this model is being applied repeatedly over time, there is a lot of redundant computation, which we can cache to increase the speed of generating a single sample.

The key insight is this – given certain nodes in the graph, we have all the information that we need to compute the current output. We call these nodes **recurrent states** by using the analogy of RNNs. These nodes have already been computed, so all we need to do is cache them on the different layers, as shown in the following diagram:

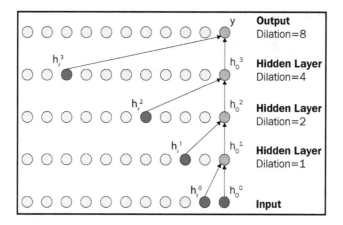

Note that at the next time point, we will need a different subset of recurrent states. As a result, we will need to cache several recurrent states per layer. The number we need to keep is equal to the dilation of that layer, as shown in the following diagram:

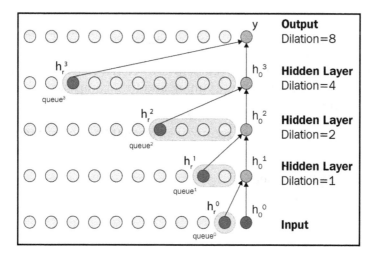

As shown in the preceding diagram with arrow marks, the number of recurrent states is the same as the dilation value in the layer.

The algorithm's components

The algorithm behind building a speech detector has two components:

- **The generation model**
- **The convolution queues**

These two components are shown in the following diagram:

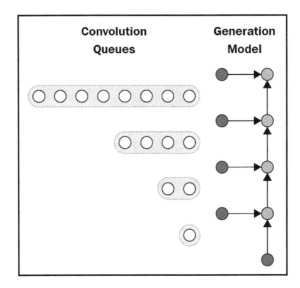

The generation model can be viewed as one step of an RNN. It takes the current observation and several recurrent states as input, and then computes the output prediction and new recurrent states.

The convolution queues store the new recurrent states that have been computed by the layer underneath it.

Let's jump into building the model.

Building the model

We will implement sentence-level English speech recognition using DeepMind's WaveNet. However, we need to consider a number of data points before building the model:

- First, while the paper on WaveNet (provided at the beginning of this chapter) used the TIMIT dataset for the speech recognition experiment, we will use the free VCTK dataset instead.

- Second, the paper added a mean pooling layer after the dilated convolution layer for downsampling. We have extracted **mel-frequency cepstral coefficients** (**MFCC**) from the `.wav` files and removed the final mean pooling layer because the original setting is impossible to run on our TitanX **Graphics Processing Unit GPU**).

- Third, since the TIMIT dataset has phoneme labels, the paper trained the model with two loss terms: **phoneme classification** and **next phoneme prediction**. Instead, we will use a single **connectionist temporal classification** (**CTC**) loss because VCTK provides sentence-level labels. As a result, we only use dilated Conv1D layers without any dilated Conv1D layers.

- Finally, we won't do quantitative analyses, such as the **bilingual evaluation understudy score** (**BLEU**) score and postprocessing by combining a language model, due to time constraints.

Dependencies

Here is a list of all the dependency libraries that will need to be installed first:

- `tensorflow`
- `sugartensor`
- `pandas`
- `librosa`
- `scikits.audiolab`

If you have problems with the `librosa` library, you can try installing `ffmpeg` using `pip`.

Datasets

We used the VCTK, LibriSpeech, and TED-LIUM release 2 datasets. The total number of sentences in the training set are composed of the previous three datasets, which equals 240,612 sentences. The validation and test sets are built using only LibriSpeech and the TED-LIUM corpus, because the VCTK corpus does not have validation and test sets. After downloading each corpus, extract them in the `asset/data/VCTK-Corpus`, `asset/data/LibriSpeech`, and `asset/data/TEDLIUM_release2` directories.

You can find the links to these datasets here:
CSTR VCTK
Corpus: `http://homepages.inf.ed.ac.uk/jyamagis/page3/page58/page58.html`
LibriSpeech ASR corpus: `http://www.openslr.org/12`
TED-LIUM: `http://www-lium.univ-lemans.fr/en/content/ted-lium-corpus`

Preprocessing the dataset

The TED-LIUM release 2 dataset provides audio data in the SPH format, so we should convert it into a format that the `librosa` library can handle. To do this, run the following command in the `asset/data` directory to convert the SPH format into the WAV format:

```
find -type f -name '*.sph' | awk '{printf "sox -t sph %s -b 16 -t wav
%s\n", $0, $0".wav" }' | bash
```

If you don't have `sox` installed, please install it first.

We found that the main bottleneck is the disk read time when training because of the size of the audio files. It is better to have smaller audio files before processing for faster execution. So, we have decided to preprocess the whole audio data into the MFCC feature files, which are much smaller. Additionally, we highly recommend using a **solid-state drive (SSD)** instead of a hard drive.

Run the following command in the console to preprocess the whole dataset:

```
python preprocess.py
```

With the processed audio files, we can now train the network.

Training the network

We will start training the network by executing the following command:

```
python train.py ( <== Use all available GPUs )
```

If you are using a machine with CUDA enabled, use the following command:

```
CUDA_VISIBLE_DEVICES=0,1 python train.py ( <== Use only GPU 0, 1 )
```

You can see the resulting `.ckpt` files and log files in the `asset/train` directory. Launch `tensorboard--logdir asset/train/log` to monitor the training process.

We've trained this model on a 3 Nvidia 1080 Pascal GPU for 40 hours until 50 epochs were reached, and then we picked the epoch when the validation loss is at a minimum. In our case, it is epoch 40. If you can see the out-of-memory error, reduce `batch_size` in the `train.py` file from 16 to 4.

The CTC losses at each epoch are as follows:

Epoch	Train set	Valid set	Test set
20	79.541500	73.645237	83.607269
30	72.884180	69.738348	80.145867
40	69.948266	66.834316	77.316114
50	69.127240	67.639895	77.866674

Here, you can see the difference between the values from the training dataset and the testing dataset. The difference is largely due to the bigger volume of data in the training dataset.

Testing the network

After training the network, you can check the validation or test set CTC loss by using the following command:

```
python test.py --set train|valid|test --frac 1.0(0.01~1.0)
```

The `frac` option will be useful if you want to test only a fraction of the dataset for fast evaluation.

Transforming a speech WAV file into English text

Next, you can convert the speech WAV file into English text by executing the following command:

```
python recognize.py --file
```

This will transform a speech WAV file into an English sentence.

The result will be printed on the console; try the following command as an example:

```
python recognize.py --file asset/data/LibriSpeech/test-
clean/1089/134686/1089-134686-0000.flac
python recognize.py --file asset/data/LibriSpeech/test-
clean/1089/134686/1089-134686-0001.flac
python recognize.py --file asset/data/LibriSpeech/test-
clean/1089/134686/1089-134686-0002.flac
python recognize.py --file asset/data/LibriSpeech/test-
clean/1089/134686/1089-134686-0003.flac
python recognize.py --file asset/data/LibriSpeech/test-
clean/1089/134686/1089-134686-0004.flac
```

The result will be as follows:

he hoped there would be stoo for dinner turnips and charrats and bruzed patatos and fat mutton pieces to be ladled out in th thick peppered flower fatan sauce stuffid into you his belly counsiled him after early night fall the yetl lampse woich light hop here and there on the squalled quarter of the browfles o berty and he god in your mind numbrt tan fresh nalli is waiting on nou cold nit husband

The ground truth is as follows:

HE HOPED THERE WOULD BE STEW FOR DINNER TURNIPS AND CARROTS AND BRUISED POTATOES AND FAT MUTTON PIECES TO BE LADLED OUT IN THICK PEPPERED FLOUR FATTENED SAUCE STUFF IT INTO YOU HIS BELLY COUNSELLED HIM AFTER EARLY NIGHTFALL THE YELLOW LAMPS WOULD LIGHT UP HERE AND THERE THE SQUALID QUARTER OF THE BROTHELS HELLO BERTIE ANY GOOD IN YOUR MIND NUMBER TEN FRESH NELLY IS WAITING ON YOU GOOD NIGHT HUSBAND

As we mentioned earlier, there is no language model, so there are some cases where capital letters and punctuation are misused, or words are misspelled.

Getting the model

Unlike image problems, it's not easy to find a pretrained deep learning model for speech-to-text that gives out checkpoints. Luckily, I found the following WaveNet speech-to-text implementation. To export the model for compression, I ran the Docker image, loaded the checkpoint, and wrote it into a protocol buffers file. To run this, use the following command:

```
python export_wave_pb.py
```

We will build the graph for inference, load the checkpoint, and write it into a protocol buffer file, as follows:

```
batch_size = 1 # batch size
voca_size = data.voca_size
x = tf.placeholder(dtype=tf.sg_floatx, shape=(batch_size, None, 20))
# sequence length except zero-padding
seq_len = tf.not_equal(x.sg_sum(axis=2), 0.).sg_int().sg_sum(axis=1)
# encode audio feature
logit = get_logit(x, voca_size)
# ctc decoding
decoded, _ = tf.nn.ctc_beam_search_decoder(logit.sg_transpose(perm=[1, 0,
2]), seq_len, merge_repeated=False)
# to dense tensor
y = tf.add(tf.sparse_to_dense(decoded[0].indices, decoded[0].dense_shape,
decoded[0].values), 1, name="output")

with tf.Session() as sess:
 tf.sg_init(sess)
 saver = tf.train.Saver()
 saver.restore(sess, tf.train.latest_checkpoint('asset/train'))

graph = tf.get_default_graph()
input_graph_def = graph.as_graph_def()

with tf.Session() as sess:
 tf.sg_init(sess)
 saver = tf.train.Saver()
 saver.restore(sess, tf.train.latest_checkpoint('asset/train'))
 # Output model's graph details for reference.
 tf.train.write_graph(sess.graph_def, '/root/speech-to-text-
wavenet/asset/train', 'graph.txt', as_text=True)
 # Freeze the output graph.
 output_graph_def =
graph_util.convert_variables_to_constants(sess,input_graph_def,"output".spl
it(","))
 # Write it into .pb file.
```

```
with tfw.gfile.GFile("/root/speech-to-text-
wavenet/asset/train/wavenet_model.pb", "wb") as f:
  f.write(output_graph_def.SerializeToString())
```

Next, we need to build a TensorFlow model and quantize the model so that it can be consumed in the mobile application.

Bazel build TensorFlow and quantizing the model

To quantize the model with TensorFlow, you need to have Bazel installed and the cloned Tensorflow repository. I recommend creating a new virtual environment to install and build TensorFlow there. Once you're done, you can run the following command:

```
bazel build tensorflow/tools/graph_transforms:transform_graph
 bazel-bin/tensorflow/tools/graph_transforms/transform_graph \
 --in_graph=/your/.pb/file \
 --outputs="output_node_name" \
 --out_graph=/the/quantized/.pb/file \
 --transforms='quantize_weights'
```

You can check out the official quantization tutorial on the TensorFlow website for other options in transforms. After quantization, the model was downsized by 75%, from 15.5 MB to 4 MB due to the 8-bit conversion. Due to the time limit, I haven't calculated the letter error rate with a test set to quantify the accuracy drop before and after quantization.

For a detailed discussion on neural network quantization, there is a great post by Pete Warden, called *Neural network quantization with TensorFlow* (https://petewarden.com/2016/05/03/how-to-quantize-neural-networks-with-tensorflow/).

Note that you can also do a full 8-bit calculation graph transformation by following the instructions in this section.

The model's size is down to 5.9 MB after this conversion, and the inference time is doubled. This could be due to the fact that the 8-bit calculation is not optimized for the Intel i5 processor on the macOS platform, which was used to write the application.

So, now that we have a compressed pretrained model, let's see what else we need to deploy the model on Android.

TensorFlow ops registration

Here, we will build the TensorFlow model using Bazel to create a `.so` file that can be called by the **Java Native Interface (JNI)**, and includes all the operation libraries that we need for the pretrained WaveNet model inference. We will use built model in the Android application.

 To find out more about Bazel, you can refer the following link: `https://docs.bazel.build/versions/master/bazel-overview.html`.

Let's begin by editing the WORKSPACE file in the cloned TensorFlow repository by uncommenting and updating the paths to **Software Development Kit (SDK)** and **Native Development Kit (NDK)**.

Next, we need to find out what ops were used in the pretrained model and generate a `.so` file with that piece of information.

First, run the following command:

```
bazel build tensorflow/python/tools:print_selective_registration_header && \
 bazel-bin/tensorflow/python/tools/print_selective_registration_header \
 --graphs=path/to/graph.pb > ops_to_register.h
```

All the ops in the `.pb` file will be listed in `ops_to_register.h`.

Next, move `op_to_register.h` to `/tensorflow/tensorflow/core/framework/` and run the following command:

```
bazel build -c opt --copt="-DSELECTIVE_REGISTRATION" \
 --copt="-DSUPPORT_SELECTIVE_REGISTRATION" \
 //tensorflow/contrib/android:libtensorflow_inference.so \
 --host_crosstool_top=@bazel_tools//tools/cpp:toolchain \
 --crosstool_top=//external:android/crosstool --cpu=armeabi-v7a
```

Unfortunately, while I didn't get any error message, the `.so` file still didn't include all the ops listed in the header file:

```
Modify BUILD in /tensorflow/tensorflow/core/kernels/
```

If you haven't tried the first option and have got the list of ops in the model, you can get the ops by using the `tf.train.write_graph` command and typing the following into your Terminal:

```
grep "op: " PATH/TO/mygraph.txt | sort | uniq | sed -E
's/^.+"(.+)".?$/\1/g'
```

Next, edit the BUILD file by adding the missing ops into `android_extended_ops_group1` or `android_extended_ops_group2` in the Android libraries section. You can also make the `.so` file smaller by removing any unnecessary ops. Now, run the following command:

```
bazel build -c opt //tensorflow/contrib/android:libtensorflow_inference.so
\
  --crosstool_top=//external:android/crosstool \
  --host_crosstool_top=@bazel_tools//tools/cpp:toolchain \
  --cpu=armeabi-v7a
```

You'll find the `libtensorflow_inference.so` file, as follows:

```
bazel-bin/tensorflow/contrib/android/libtensorflow_inference.so
```

Note that while running this command on Android, we ran into an error with the `sparse_to_dense` op. If you'd like to repeat this work, add `REGISTER_KERNELS_ALL(int64);` to `sparse_to_dense_op.cc` on line 153, and compile again.

In addition to the `.so` file, we also need a JAR file. You can simply add this in the `build.gradle` file, as follows:

```
allprojects {
 repositories {
    jcenter()
    }
 }

dependencies {
 compile 'org.tensorflow:tensorflow-android:+'
 }
```

Or, you can run the following command:

```
bazel build //tensorflow/contrib/android:android_tensorflow_inference_java
```

You'll find the file, as shown in the following code block:

```
bazel-
bin/tensorflow/contrib/android/libandroid_tensorflow_inference_java.jar
```

Now, move both files into your Android project.

Building an Android application

In this section, we are going to build an Android application that will convert the user's voice input into text. Essentially, we are going to build a speech-to-text converter. We have modified the TensorFlow speech example in the TensorFlow Android demo repository for this exercise.

> You can find the TensorFlow Android demo application at `https://github.com/tensorflow/tensorflow/tree/master/tensorflow/examples/android`.

The `build.gradle` file in the demo actually helps you build the `.so` and JAR files. So, if you'd like to start the demo examples with your own model, you can simply get the list of your ops, modify the BUILD file, and let the `build.gradle` file take care of the rest. We will get into the details of setting up the Android application in the following sections.

Requirements

The requirements you will need to build the Android application are as follows:

- TensorFlow 1.13
- Python 3.7
- NumPy 1.15
- python-speech-features

> TensorFlow link: `https://github.com/tensorflow/tensorflow/releases`
> Python link: `https://pip.pypa.io/en/stable/installing/`
> Numpy
> link: `https://docs.scipy.org/doc/numpy-1.13.0/user/install.html`
> Python-speech-features link: `https://github.com/jameslyons/python_speech_features`

Now, let's start building the Android application from scratch. In this application, we will record audio and then convert it into text.

> Set up Android Studio based on your operating system by going to the following link: `https://developer.android.com/studio/install`. The code repository that's used in this project has been modified from the TensorFlow example provided here: `https://github.com/tensorflow/tensorflow/tree/master/tensorflow/examples/android`.

We will use the TensorFlow sample application and edit it according to our needs.

Add **Application name** and the **Company domain** name, as shown in the following screenshot:

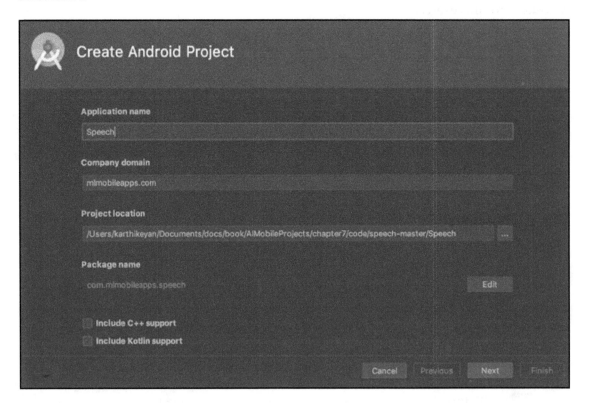

In the next step, select the **Target Android Devices** version. We will select the minimum version as **API 15**:

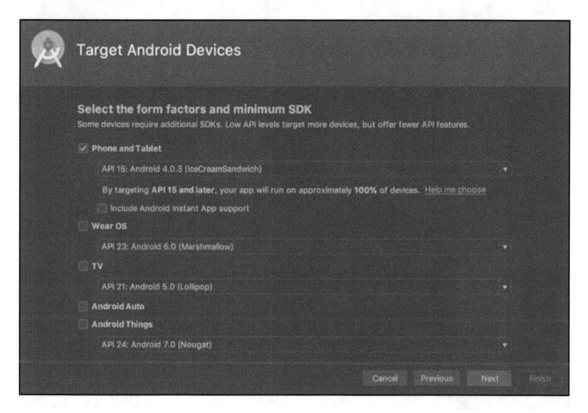

After this, we will add either **Empty Activity** or **No Activity**:

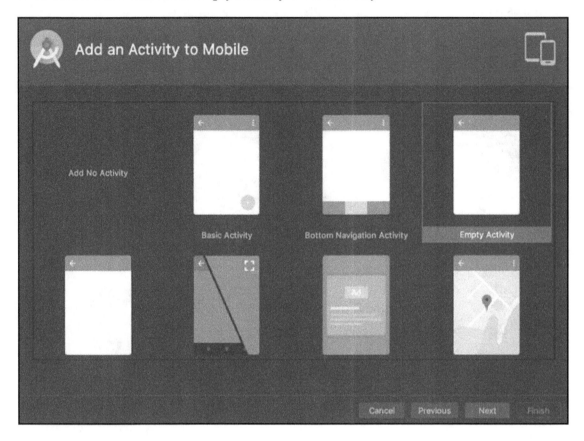

Now, let's start adding the activity and use the generated TensorFlow model to get the result. We need to enable two permissions so that we can use them in our application, as shown in the following code block:

```
<manifest xmlns:android="http://schemas.android.com/apk/res/android"
    package="com.mlmobileapps.speech">

    <uses-permission
android:name="android.permission.WRITE_EXTERNAL_STORAGE"/>
    <uses-permission android:name="android.permission.RECORD_AUDIO" />

    <uses-sdk
        android:minSdkVersion="25"
        android:targetSdkVersion="25" />

    <application android:allowBackup="true"
```

```
            android:debuggable="true"
            android:label="@string/app_name"
            android:icon="@drawable/ic_launcher"
            android:theme="@style/MaterialTheme">

            <activity android:name="org.tensorflow.demo.SpeechActivity"
                android:screenOrientation="portrait"
                android:label="@string/activity_name_speech">
                <intent-filter>
                    <action android:name="android.intent.action.MAIN" />
                    <category android:name="android.intent.category.LAUNCHER" />
                </intent-filter>
            </activity>
        </application>
    </manifest>
```

We will have a minimal UI for the application, with a couple of `TextView` components and a `Button`:

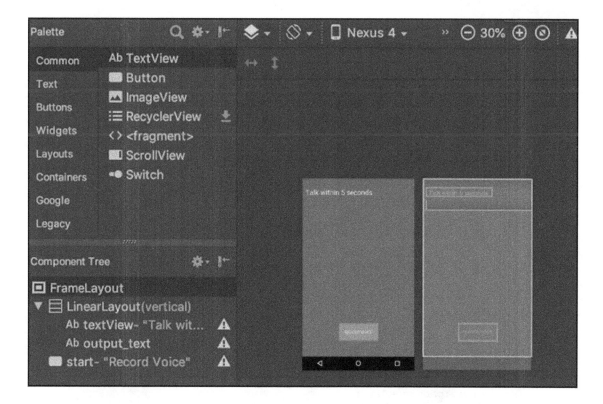

The following XML layout mimics the UI in the preceding screenshot:

```xml
<FrameLayout
    xmlns:android="http://schemas.android.com/apk/res/android"
    xmlns:app="http://schemas.android.com/apk/res-auto"
    xmlns:tools="http://schemas.android.com/tools"
    android:layout_width="match_parent"
    android:layout_height="match_parent"
    android:background="#6200EE"
    tools:context="org.tensorflow.demo.SpeechActivity">

    <LinearLayout
        android:layout_width="match_parent"
        android:orientation="vertical"
        android:layout_height="wrap_content">
        <TextView
            android:id="@+id/textView"
            android:layout_width="wrap_content"
            android:layout_height="wrap_content"

            android:layout_gravity="top"
            android:textColor="#fff"
            android:layout_marginLeft="10dp"
            android:layout_marginTop="30dp"
            android:text="Talk within 5 seconds"
            android:textAlignment="center"
            android:textSize="24dp" />

        <TextView
            android:id="@+id/output_text"
            android:layout_width="wrap_content"
            android:layout_height="wrap_content"
            android:textColor="#fff"
            android:layout_gravity="top"
            android:layout_marginLeft="10dp"
            android:layout_marginTop="10dp"
            android:textAlignment="center"
            android:textSize="24dp" />
    </LinearLayout>

    <Button
        android:id="@+id/start"
        android:background="#ff0266"
        android:textColor="#fff"
        android:layout_width="wrap_content"
        android:padding="20dp"
        android:layout_height="wrap_content"
        android:layout_gravity="bottom|center_horizontal"
```

```
            android:layout_marginBottom="50dp"
            android:text="Record Voice" />

    </FrameLayout>
```

Let's add the steps for the speech recognizer activity, as follows:

```
@Override
protected void onCreate(Bundle savedInstanceState) {
    // Set up the UI.
    super.onCreate(savedInstanceState);
    setContentView(R.layout.activity_speech);
    startButton = (Button) findViewById(R.id.start);
    startButton.setOnClickListener(
        new View.OnClickListener() {
            @Override
            public void onClick(View view) {
                startRecording();
            }
        });
    outputText = (TextView) findViewById(R.id.output_text);
    // Load the Pretrained WaveNet model.
    inferenceInterface = new TensorFlowInferenceInterface(getAssets(),
MODEL_FILENAME);
    requestMicrophonePermission();
}
```

 Note that we are not going to discuss the basics of Android here.

Next, we will launch the recorder, as follows:

```
public synchronized void startRecording() {
    if (recordingThread != null) {
        return;
    }
    shouldContinue = true;
    recordingThread =
        new Thread(
            new Runnable() {
                @Override
                public void run() {
                    record();
                }
```

```
        });
    recordingThread.start();
}
```

The following code shows the implementation of the `record()` method:

```
private void record() {
android.os.Process.setThreadPriority(android.os.Process.THREAD_PRIORITY_AUD
IO);

  // Estimate the buffer size we'll need for this device.
  int bufferSize =
      AudioRecord.getMinBufferSize(
              SAMPLE_RATE, AudioFormat.CHANNEL_IN_MONO,
AudioFormat.ENCODING_PCM_16BIT);
  if (bufferSize == AudioRecord.ERROR || bufferSize ==
AudioRecord.ERROR_BAD_VALUE) {
    bufferSize = SAMPLE_RATE * 2;
  }
  short[] audioBuffer = new short[bufferSize / 2];

  AudioRecord record =
      new AudioRecord(
          MediaRecorder.AudioSource.DEFAULT,
          SAMPLE_RATE,
          AudioFormat.CHANNEL_IN_MONO,
          AudioFormat.ENCODING_PCM_16BIT,
          bufferSize);

  if (record.getState() != AudioRecord.STATE_INITIALIZED) {
    Log.e(LOG_TAG, "Audio Record can't initialize!");
    return;
  }

  record.startRecording();

  Log.v(LOG_TAG, "Start recording");

  while (shouldContinue) {
    int numberRead = record.read(audioBuffer, 0, audioBuffer.length);
      Log.v(LOG_TAG, "read: " + numberRead);
    int maxLength = recordingBuffer.length;
    recordingBufferLock.lock();
    try {
        if (recordingOffset + numberRead < maxLength) {
            System.arraycopy(audioBuffer, 0, recordingBuffer,
recordingOffset, numberRead);
        } else {
```

```
            shouldContinue = false;
        }
        recordingOffset += numberRead;
    } finally {
        recordingBufferLock.unlock();
    }
}
record.stop();
record.release();
startRecognition();
}
```

The following code shows the implementation of the audio recognizing method:

```
public synchronized void startRecognition() {
    if (recognitionThread != null) {
        return;
    }
    shouldContinueRecognition = true;
    recognitionThread =
        new Thread(
            new Runnable() {
                @Override
                public void run() {
                    recognize();
                }
            });
    recognitionThread.start();
}

private void recognize() {
    Log.v(LOG_TAG, "Start recognition");

    short[] inputBuffer = new short[RECORDING_LENGTH];
    double[] doubleInputBuffer = new double[RECORDING_LENGTH];
    long[] outputScores = new long[157];
    String[] outputScoresNames = new String[]{OUTPUT_SCORES_NAME};


        recordingBufferLock.lock();
        try {
          int maxLength = recordingBuffer.length;
            System.arraycopy(recordingBuffer, 0, inputBuffer, 0, maxLength);
        } finally {
          recordingBufferLock.unlock();
        }

        // We need to feed in float values between -1.0 and 1.0, so divide the
```

```
  // signed 16-bit inputs.
  for (int i = 0; i < RECORDING_LENGTH; ++i) {
    doubleInputBuffer[i] = inputBuffer[i] / 32767.0;
  }

  //MFCC java library.
  MFCC mfccConvert = new MFCC();
  float[] mfccInput = mfccConvert.process(doubleInputBuffer);
  Log.v(LOG_TAG, "MFCC Input======> " + Arrays.toString(mfccInput));

  // Run the model.
  inferenceInterface.feed(INPUT_DATA_NAME, mfccInput, 1, 157, 20);
  inferenceInterface.run(outputScoresNames);
  inferenceInterface.fetch(OUTPUT_SCORES_NAME, outputScores);
  Log.v(LOG_TAG, "OUTPUT======> " + Arrays.toString(outputScores));


  //Output the result.
  String result = "";
  for (int i = 0;i<outputScores.length;i++) {
      if (outputScores[i] == 0)
          break;
      result += map[(int) outputScores[i]];
  }
  final String r = result;
  this.runOnUiThread(new Runnable() {
      @Override
      public void run() {
          outputText.setText(r);
      }
  });
  Log.v(LOG_TAG, "End recognition: " +result);
}
```

The model is run through the `TensorFlowInferenceInterface` class, as shown in the preceding code.

Once we have the completed code running, run the application.

On the first run, you will need to allow the application to use the phone's internal microphone, as demonstrated in the following screenshot:

Once we give permission to use the microphone, click on **RECORD VOICE** and give your voice input within **5 seconds**. There are two attempts shown in the following screenshots for the `how are you` input keyword with an Indian accent. It works better with US and UK accents.

The first attempt is as follows:

The second attempt is as follows:

You should try this with your own accent to get the correct output. This is a very simple way to start building your own speech detector that you can improve on even further.

Summary

In this chapter, you learned how to build a complete speech detector on your own. We discussed how the WaveNet model works in detail. With this application, we can make a simple speech-to-text converter work; however, a lot of improvements and updates need to be done to get perfect results. You can build the same application on the iOS platform as well by converting the model into CoreML.

In the next chapter, we will move on and build a handwritten digit classifier using the **Modified National Institute of Standards and Technology (MNIST)** model.

8
Implementing GANs to Recognize Handwritten Digits

In this chapter, we will build an Android application that detects handwritten numbers and works out what the number is by using adversarial learning. We will use the **Modified National Institute of Standards and Technology (MNIST)** dataset for digit classification. We will also look into the basics of **Generative Adversarial Networks (GANs)**.

In this chapter, we will take a closer look at the following topics:

- Introduction to GANs
- Understanding the MNIST database
- Building the TensorFlow model
- Building the Android application

 The code for this application can be found at `https://github.com/intrepidkarthi/AImobileapps`.

Introduction to GANs

GANs are a class of **machine learning (ML)** algorithm that's used in unsupervised ML. They are comprised of two deep neural networks that are competing against each other (so it is termed as adversarial). GANs were introduced at the University of Montreal in 2014 by Ian Goodfellow and other researchers, including Yoshua Bengio.

Ian Goodfellow's paper on GANs can be found at `https://arxiv.org/abs/1406.2661`.

GANs have the potential to mimic any data. This means that GANs can be trained to create similar versions of any data, such as images, audio, or text. A simple workflow of a GAN is shown in the following diagram:

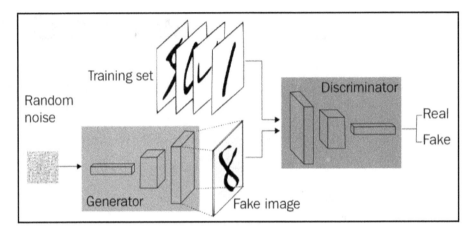

The workflow of the GAN will be explained in the following sections.

Generative versus discriminative algorithms

To understand GANs, we must know how discriminative and generative algorithms work. Discriminative algorithms try to predict a label and classify the input data, or categorize them to where the data belongs. On the other hand, generative algorithms make an attempt to predict features to give a certain label.

For example, a discriminative algorithm can predict whether an email is spam or not. Here, spam is one of the labels, and the text that's captured from the message is considered the input data. If you consider the label as y and the input as x, we can formulate this as follows:

$$p(y|x)$$

On the other hand, generative algorithms try to guess how likely these input features (x, in the previous equation) are. Generative models care about how you get x, while discriminative models care about the relation between x and y.

Using the MNIST database as an example, the generator will generate images and pass them on to the discriminator. The discriminator will authenticate the image if it is truly from the MNIST dataset. The generator generates images with the hope that it will pass through the discriminator and be authenticated, even though it is fake (as shown in the preceding diagram).

How GANs work

Based on our example, we will assume that we are passing numbers as inputs:

1. The generator takes random numbers as inputs and returns an image as the output
2. The output image is passed into the discriminator, and, at the same time, the discriminator receives input from the dataset
3. The discriminator takes in both real and fake input images, and returns probabilities between zero and one (with one representing a prediction of authenticity and zero representing a fake)

Using the example application we discussed in this chapter, we can use the same steps to pass the user's hand-drawn image as one of the fake images and try to find the probability value of it being correct.

Understanding the MNIST database

The MNIST dataset consists of 60,000 handwritten digits. It also consists of a test dataset made up of 10,000 digits. While it is a subset of the NIST dataset, all the digits in this dataset are size normalized and have been centered on a 28 x 28 pixels sized image. Here, every pixel contains a value of 0-255 with its grayscale value.

The MNIST dataset can be found at `http://yann.lecun.com/exdb/mnist/`. The NIST dataset can be found at `https://www.nist.gov/srd/nist-special-database-19`.

Building the TensorFlow model

In this application, we will build an MNIST dataset based TensorFlow model that we will use in our Android application. Once we have the TensorFlow model, we will convert it into a TensorFlow Lite model. The step-by-step procedure of downloading the model and building the TensorFlow model is as follows.

Here is the architecture diagram on how our model works. The way to achieve this is explained as follows:

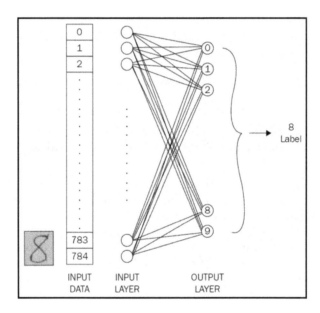

Using TensorFlow, we can download the MNIST data with one line of Python code, as follows:

```
import tensorflow as tf
from tensorflow.examples.tutorials.mnist import input_data
#Reading data
mnist = input_data.read_data_sets("./data/", one_hot-True)
```

Now, we have the MNIST dataset downloaded. After that, we will read the data, as shown in the previous code.

Now, we can run the script to download the dataset. We will run the script from the console, as follows:

```
> python mnist.py
Successfully downloaded train-images-idx3-ubyte.gz 9912422 bytes.
Extracting MNIST_data/train-images-idx3-ubyte.gz Successfully downloaded
train-labels-idx1-ubyte.gz 28881 bytes.
Extracting MNIST_data/train -labels -idx1 -ubyte.gz
Successfully downloaded t10k -images -idx3 -ubyte.gz 1648877 bytes.
Extracting MNIST_data/t10k -images -idx3 -ubyte.gz
Successfully downloaded t10k -labels -idx1 -ubyte.gz 4542 bytes. Extracting
MNIST_data/t10k -labels -idx1 -ubyte.gz
```

Once we have the dataset ready, we will add a few variables that we will use in our application, as follows:

```
image_size = 28
labels_size = 10
learning_rate = 0.05
steps_number = 1000
batch size = 100
```

We need to define these variables to control the parameters on building the model as required by the TensorFlow framework. This classification process is simple. The number of pixels that exist in a 28 x 28 image is 784. So, we have a corresponding number of input layers. Once we have the architecture set up, we will train the network and evaluate the results, obtained to understand the effectiveness and accuracy of the model.

Now, let's define the variables that we added in the preceding code block. Depending on whether the model is in the training phase or the testing phase, different data will be passed through the classifier. The training process needs labels to be able to match them to current predictions. This is defined in the following variable:

```
#Define placeholders
training_data = tf.placeholder(tf.float32, [None, image_size*image_size])
labels = tf.placeholder(tf.float32, [None, labels_size])
```

As the computation-graph evaluation occurs, placeholders will be filled. In the training process, we adjust the values of biases and weights toward increasing the accuracy of our results. To achieve this, we will define the weight and bias parameters, as follows:

```
#Variables to be tuned
W = tf.Variable(tf.truncated_normal([image_size*image_size, labels_size],
stddev=0.1))
b = tf.Variable(tf.constant(0.1, shape-[labels_size]))
```

Once we have variables that can be tuned, we can move on to building the output layer in just one step:

```
#Build the network
output = tf.matmul(training_data, W) + b
```

We have successfully built the output layer of the network with the training data.

Training the neural network

By optimizing loss, we can get the training process to work. We need to reduce the difference between the actual label value and the network prediction. The term to define this loss is **cross entropy**.

In TensorFlow, cross entropy is provided by the following method:

```
tf.nn.softmax_cross_entropy_with_logits
```

This method applies softmax to the model's prediction. Softmax is similar to logistic regression, and produces a decimal between 0 and 1.0. For example, a logistic regression output of 0.9 from an email classifier suggests a 90% chance of an email being spam and a 10% chance of it not being spam. The sum of all the probabilities is 1.0, as shown with an example in the following table:

Object	Probability
Apple	0.05
Car	0.80
Sunflower	0.01
Cup	0.14

Softmax is implemented through a neural network layer, just before the output layer. The softmax layer must have the same number of nodes as the output layer.

Loss is defined using the `tf.reduce_mean` method, and the `GradientDescentOptimizer()` method is used in training steps to minimize the loss. This is shown in the following code:

```
#Defining the loss
loss = tf.reduce_mean(tf.nn.softmax_cross_entropy_with_logits(labels-
labels, logits-output))
# Training step with gradient descent
```

```
train_step =
tf.train.GradientDescentOptimizer(learning_rate).minimize(loss)
```

The `GradientDescentOptimizer` method will take several steps by adjusting the values of *w* and *b* (the weight and bias parameters) in the output. The values will be adjusted until we reduce loss and are closer to a more accurate prediction, as follows:

```
# Accuracy calculation
correct_prediction = tf.equal(tf.argmax(output, 1), tf.argmax(labels, 1))
accuracy = tf.reduce_mean(tf.cast(correct_prediction, tf.float32))
```

We start the training by initializing the session and the variables, as follows:

```
# Run the training
sess = tf.InteractiveSession() sess.run(tf.global_variables_initializer())
```

Based on the parameters of the number of steps (`steps_number`) defined previously, the algorithm will run in a loop. We will then run the optimizer, as follows:

```
for i in range(steps_number):
    # Get the next batch input_batch,
    labels_batch = mnist.train.next_batch(batch_size)
    feed_dict = {training_data: input_batch, labels: labels_batch}
    # Run the training step
    train_step.run(feed_dict=feed_dict)
```

With TensorFlow, we can measure the accuracy of our algorithm and print the accuracy value. We can keep it improving as long as the accuracy level increases and finds the threshold value on where to stop, as follows:

```
# Print the accuracy progress on the batch every 100 steps
if i%100 == 0:
    train_accuracy = accuracy.eval(feed_dict=feed_dict)
    print("Step %d, batch accuracy %g %%"%(i, train_accuracy*100))
```

Once the training is done, we can evaluate the network's performance. We can use the training data to measure performance, as follows:

```
# Evaluate on the test set
test_accuracy = accuracy.eval(feed_dict=ftraining_data: mnist.test.images,
labels: mnist.test.labels})
print("Test accuracy: %g %%"%(test_accuracy*100))
```

When we run the Python script, the output on the console is as follows:

```
Step 0, training batch accuracy 13 %
Step 100, training batch accuracy 80 %
Step 200, training batch accuracy 87 %
Step 300, training batch accuracy 81 %
Step 400, training batch accuracy 86 %
Step 500, training batch accuracy 85 %
Step 600, training batch accuracy 89 %
Step 700, training batch accuracy 90 %
Step 800, training batch accuracy 94 %
Step 900, training batch accuracy: 89.49 %
Test accuracy 91 %
```

Now, we have arrived at an accuracy rate of 89.2%. When we try to optimize our results more, the accuracy level reduces; this is where we set have our threshold value to stop the training.

Let's build the TensorFlow model for the MNIST dataset. Inside the TensorFlow framework, the scripts that are provided save the MNIST dataset into a TensorFlow (.pb) model. The same script is attached to this application's repository.

 The code for this application can be found at `https://github.com/intrepidkarthi/AImobileapps`.

We will begin by training the model using the following Python code line:

```
$:python mnist.py
```

We will now run the script to generate our model.

The following script helps us export the model by adding some additional parameters:

```
python mnist.py --export_dir /./mnist_model
```

The saved model can be found in the time stamped directory under `/./mnist_model/` (for example, `/./mnist_model/1536628294/`).

The obtained TensorFlow model will be converted into a TensorFlow Lite model using `toco`, as follows:

```
toco \
--input_format=TENSORFLOW_GRAPHDEF
--output_format=TFLITE \
--output_file=./mnist.tflite \
```

```
--inference_type=FLOAT \
--input_type=FLOAT
--input_arrays=x \
--output_arrays=output
--input_shapes=1,28,28,1 \
--graph_def_file=./mnist.pb
```

Toco is a command-line tool that's used to run the **TensorFlow Lite Optimizing Converter** (**TOCO**), which converts a TensorFlow model into a TensorFlow Lite model. The preceding `toco` command produces `mnist.tflite` as its output, which we will use in our application in the next section.

Building the Android application

Let's create the Android application step-by-step with the model that we have built. We will start by creating a new project in Android Studio:

1. Create a new application in Android Studio:

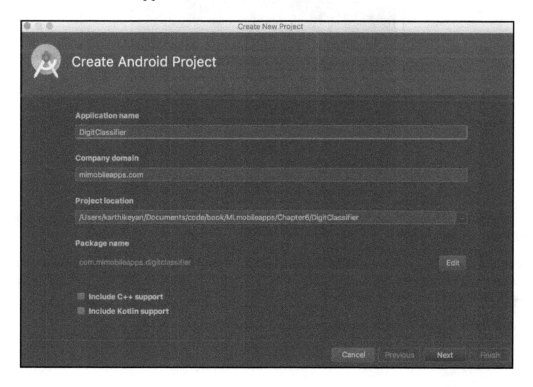

2. Drag the created TensorFlow Lite model to the `assets` folder, along with the `labels.txt` file. We will read the model and label from the assets folder:

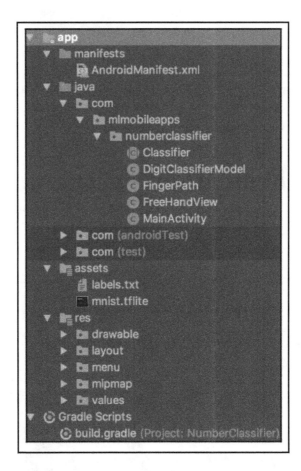

The preceding screenshot shows the file structure in the project. If necessary, we can store the model file inside the secondary memory storage as well.

One of the advantages of **FreeHandView** is that we can create a simple view where users can draw any number of digits. In addition to this, the bar chart on the screen will show the classification of the detected number.

We will use a step-by-step procedure to create the classifier.

Here is the `FreeHandView` constructor method that we will use to draw the digits. We initialize the `Paint` object with the necessary parameters, as follows:

```
public FreeHandView(Context context, AttributeSet attrs) {
  super(context, attrs);
  mPaint = new Paint();
  mPaint.setAntiAlias(true);
  mPaint.setDither(true);
  mPaint.setColor(DEFAULT_COLOR);
  mPaint.setStyle(Paint.Style.STROKE);
  mPaint.setStrokeJoin(Paint.Join.ROUND);
  mPaint.setStrokeCap(Paint.Cap.ROUND);
  mPaint.setXfermode(null); mPaint.setAlpha(0xff);
  mEmboss = new EmbossMaskFilter(new float[] I1, 1, 1}, 0.4f, 6, 3.5f);
  mBlur = new BlurMaskFilter(5, BlurMaskFilter.Blur.NORMAL);
}
```

The functions of each parameter that was used in the preceding code block are explained as follows:

- `mPaint.setAntiAlias(true)`: A helper for `setFlags()`, setting or clearing the `ANTI_ALIAS_FLAG` bit. Antialiasing smooths out the edges of what is being drawn, but it has no impact on the interior of the shape.
- `mPaint.setDither(true)`: A helper for `setFlags()`, setting or clearing the `DITHER_FLAG` bit. Dithering affects how colors that are higher precision than the device are down-sampled.
- `mPaint.setColor(DEFAULT_COLOR)`: Sets the paint's color.
- `mPaint.setStyle(Paint.Style.STROKE)`: Sets the paint's style, used for controlling how primitives' geometries are interpreted (except for `drawBitmap`, which always assumes `Fill`).
- `mPaint.setStrokeJoin(Paint.Join.ROUND)`: Sets the paint's `Join`.
- `mPaint.setStrokeCap(Paint.Cap.ROUND)`: Sets the paint's `Cap`.
- `mPaint.setXfermode(null)`: Sets or clears the transfer mode object.
- `mPaint.setAlpha(0xff)`: A helper to `setColor()`, that only assigns the color's `alpha` value, leaving its `r`, `g`, and `b` values unchanged.

Inside the `init()` method of the view life cycle, we will initialize the `ImageClassifier`, and pass on the `BarChart` object:

```
public void init(DisplayMetrics metrics, ImageClassifier classifier,
BarChart barChart) {
  int height = 1000;
  int width = 1000;
```

```
mBitmap = Bitmap.createBitmap(width, height, Bitmap.Config.ARGB_8888);
mCanvas = new Canvas(mBitmap);
currentColor = DEFAULT_COLOR; strokeWidth = BRUSH_SIZE;
mClassifier = classifier;
this.predictionBar = predictionBar;
this.barChart = barChart; addValuesToBarEntryLabels();
}
```

 We will use the chart from the following library: `https://github.com/PhilJay/MPAndroidChart`.

We will initialize the `BarChart` view, with the *x* axis containing numbers from zero to nine and the *y* axis containing the probability value from 0 to 1.0:

```
BarChart barChart = (BarChart) findViewByld(R.id.barChart);
 barChart.animateY(3000);
 barChart.getXAxis().setEnabled(true);
 barChart.getAxisRight().setEnabled(false);
 barChart.getAxisLeft().setAxisMinimum(0.0f); // start at zero
 barChart.getAxisLeft().setAxisMaximum(1.0f); // the axis maximum is 100
 barChart.getDescription().setEnabled(false);
 barChart.getLegend().setEnabled(false);
 // the labels that should be drawn on the X-Axis final String[] barLabels =
 new String[]{"0", "1", "2", "3", "4", "5", "6", n7,1, 118n, n9,1};

 //To format the value as integers
 IAxisValueFormatter formatter = new IAxisValueFormatter() {
 @Override public String getFormattedValue(float value, AxisBase axis) {
  return barLabels[(int) value); }
 };
 barChart.getXAxis().setGranularity(0f); // minimum axis-step (interval) is
1
 barChart.getXAxis().setValueFormatter(formatter);
 barChart.getXAxis().setPosition(XAxis.XAxisPosition.BOTTOM);
 barChart.getXAxis().setTextSize(5f);
```

Once we have initialized the `BarChart` view, we will call the `OnDraw()` method of the view life cycle, which applies strokes in accordance with the path of the user's finger movements. The `OnDraw()` method is called as part of the view life cycle method once the `BarChart` view is initialized.

Inside the `OnDraw` method, we will track the finger movement of the user, and the same movements will be drawn on the canvas, as follows:

```
@Override protected void onDraw(Canvas canvas) {
```

```
canvas.save();
mCanvas.drawColor(backgroundColor);
for (FingerPath fp : paths) {
mPaint.setColor(fp.color);
mPaint.setStrokeWidth(fp.strokeWidth);
mPaint.setMaskFilter(null);
    if (fp.emboss)
     mPaint.setMaskFilter(mEmboss);
    else if (fp.blur)
     mPaint.setMaskFilter(mBlur);
    mCanvas.drawPath(fp.path, mPaint);
}
canvas.drawBitmap(mBitmap, 0, 0, mBitmapPaint); canvas. restore();
}
```

Inside the `onTouchEvent()` method, we can track the user's finger position using the move, up, and down events and initiate actions based upon that. This is one of the methods in the view's life cycle that's used to track events. There are three events that will be triggered when you touch your mobile based on finger movements. In the case of `action_down` and `action_move`, we will handle events to draw the on-hand movement on the view with the initial paint object attributes. When the `action_up` event is triggered, we will save the view into a file, as well as pass the file image to the classifier to identify the digit. After that, we will represent the probability values using the `BarChart` view. These steps are as follows:

```
@Override public boolean onTouchEvent(MotionEvent event) {
  float x = event.getX();
  float y = event.getY();
  BarData exampleData;
  switch(event.getAction()) {
      case MotionEvent.ACTION_DOWN :
      touchStart(x, y);
      invalidate();
      break;
      case MotionEvent.ACTION_MOVE :
      touchMove(x, y);
      invalidate();
      break;
      case MotionEvent.ACTION_UP :
      touchUp();
    Bitmap scaledBitmap = Bitmap.createScaledBitmap(mBitmap,
mClassifier.getImageSizeX(), mClassifier.getImageSizeY(), true);
      Random rng = new Random();
      try {
      File mFile;
      mFile = this.getContext().getExternalFilesDir(String.valueOf
(rng.nextLong() + ".png"));
```

```
        FileOutputStream pngFile = new FileOutputStream(mFile);
        }
    catch (Exception e){ }
    //scaledBitmap.compress(Bitmap.CompressFormat.PNG, 90, pngFile);
    Float prediction = mClassifier.classifyFrame(scaledBitmap);
    exampleData = updateBarEntry();
    barChart.animateY(1000, Easing.EasingOption.EaseOutQuad);
    XAxis xAxis = barChart.getXAxis();
    xAxis.setValueFormatter(new IAxisValueFormatter() {
    @Override public String getFormattedValue(float value, AxisBase axis)
{
        return xAxisLabel.get((int) value);
    });
    barChart.setData(exampleData);
    exampleData.notifyDataChanged(); // let the data know a // dataset
changed
    barChart.notifyDataSetChanged(); // let the chart know it's // data
changed
    break;
    }
 return true;
}
```

Inside the `ACTION_UP` action, there is a `updateBarEntry()` method call. This is where we call the classifier to get the probability of the result. This method also updates the `BarChart` view based on the results from the classifier, as follows:

```
public BarData updateBarEntry() {
    ArrayList<BarEntry> mBarEntry = new ArrayList<>();
    for (int j = 0; j < 10; ++j) {
        mBarEntry.add(new BarEntry(j, mClassifier.getProbability(j)));
    }
    BarDataSet mBarDataSet = new BarDataSet(mBarEntry, "Projects");
    mBarDataSet.setColors(ColorTemplate.COLORFUL_COLORS);
    BarData mBardData = new BarData(mBarDataSet);
    return mBardData;
}
```

FreeHandView looks like this, along with an empty bar chart:

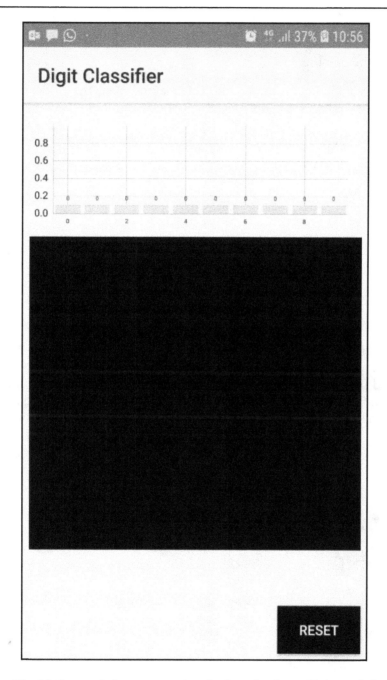

With this, we will add the module to recognize the handwritten digits and then classify them.

Digit classifier

Now, let's write the classifier.

1. First, we will load the model file. This method reads the model from the assets folder and loads it into the memory:

```
/** Memory-map the model file in Assets. */
private MappedByteBuffer loadModelFile(Activity activity) throws
IOException
{
    AssetFileDescriptor fileDescriptor =
activity.getAssets().openFd(getModelPath());
    FileInputStream inputStream = new
FileInputStream(fileDescriptor.getFileDescriptor());
    FileChannel fileChannel = inputStream.getChannel();
    long startOffset = fileDescriptor.getStartOffset();
    long declaredLength = fileDescriptor.getDeclaredLength();
return fileChannel.map(FileChannel.MapMode.READ_ONLY, startOffset,
declaredLength);
}
```

2. Now, let's write the TensorFlow Lite classifier, frame-by-frame. This is the place where we get the results from the digit classifier. Once we have received the saved file image as the user input, the bitmap will be converted into a byte buffer to run the inference on top of the model. Once we have received the output, the time taken to get the results are noted using the SystemClock time:

```
/** Classifies a frame from the preview stream. */
public float classifyFrame(Bitmap bitmap)
{
    if (tflite == null){
    Log.e(TAG, "classifier has not been initialized; Skipped.");
return 0.5f;
    }
convertBitmapToByteBuffer(bitmap); // Here's where the
classification happens!!!
long startTime = SystemClock.uptimeMillis();
runInference();
long endTime = SystemClock.uptimeMillis();
Log.d(TAG, "Timecost to run model inference: " +
Long.toString(endTime - startTime));
return getProbability(0);
}
```

3. The `runlnference()` method calls the `run` method from `tflite`, as follows:

```
@Override
protected void runlnference()
{
    tflite.run(imgData, labelProbArray);
}
```

4. Next, let's start the application from `MainActivity`, where the `barChart` view is initialized. Initialize the `barChart` view on the *x* and *y* axis, along with the following values:

```
BARENTRY = new ArrayList<>(); initializeBARENTRY();
Bardataset = new BarDataSet(BARENTRY, "project");
BARDATA = new BarData(Bardataset);
barChart.setData(BARDATA);
```

5. Initialize FreeHandView to start classifying inside the `OnCreate()` method of `MainActivity`:

```
paintView.init(metrics, classifier, barChart);
```

6. When you reach the probability value of 1.00, the algorithm identifies the digit with 100% accuracy. An example of this is shown here:

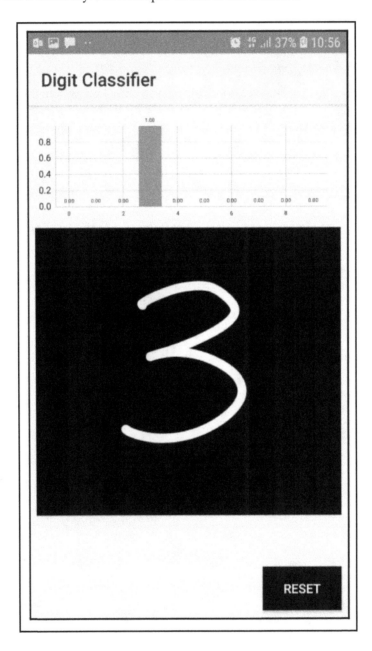

7. There are instances in which the classification decreases the probability with partial matches, as shown in the following screenshot:

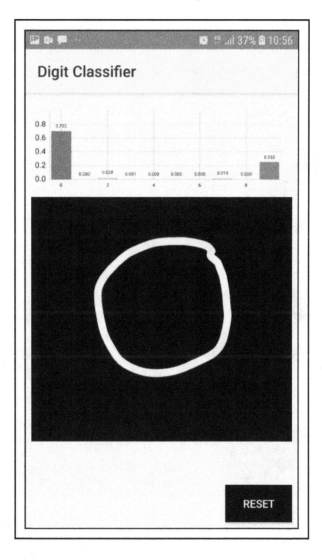

8. There are also other instances where the probability ends up with multiple partial matches. An example of this is shown in the following screenshot:

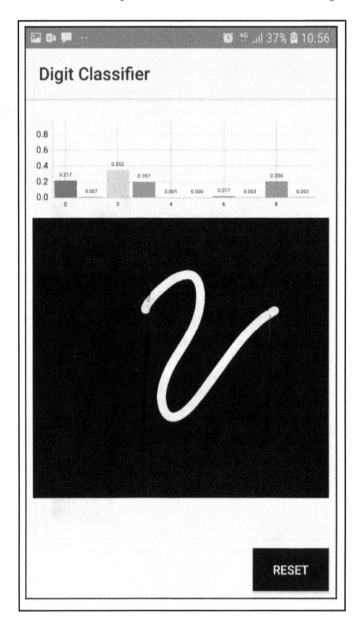

Any such situation requires more rigorous training of the model.

9. Clicking on the **RESET** button will clear up the view so that you can draw again. We will implement it using the following lines of code:

```
resetButton.setOnClickListener(new View.OnClickListener() {
 public void onClick(View v) {
 paintView.clear();
 }
}) ;
```

Once you click on the **RESET** button, the preceding code clears up the FreeHandView area, as follows:

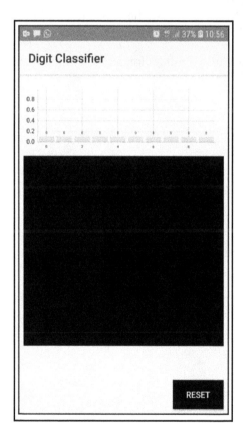

You can also check that the application works properly by writing characters other than digits, and checking the performance of the output on the bar chart.

In this section, we learned how the application classifies the different digits that are hand-drawn, and also provides the probability of those digits being correct.

Summary

Using this Android application, we can learn how to write a freehand writing classifier using TensorFlow Lite. With more data on handwritten alphabet datasets, we should be able to identify alphabets in any language using GANs.

In the next chapter, we will build a model for sentiment analysis and build an app on top of it.

Sentiment Analysis over Text Using LinearSVC

9

In this chapter, we are going to build an iOS application to do sentiment analysis over text and image through user input. We will use existing data models that were built for the same purpose by using LinearSVC, and convert those models into core **machine learning (ML)** models for ease of use in our application.

Sentiment analysis is the process of identifying a feeling or opinion that is inspired by any given data in the form of text, image, audio, or video. There are a lot of use cases for sentiment analysis. Even now, political parties can easily identify the general mindset of the people who are going to elect them and they also have the potential to change that mindset.

Let's take a look at building our own ML model on sentiment analysis from an existing dataset. In this chapter, we will look at the following topics:

- Building the ML model using scikit-learn
- **Linear Support Vector Classification** (**LinearSVC**)
- Building an iOS application

Building the ML model using scikit–learn

In this section, we will build our own model. There are existing datasets available that are related to Twitter feed data on the topic of product and movie reviews. You can pick a dataset that suits you; in this chapter, we will pick a dataset that has customer reviews.

A dataset that contains both positive and negative reviews of customers can be found at `http://boston.lti.cs.cmu.edu/classes/95-865-K/HW/HW3/`. You can download the dataset from the following link: `http://boston.lti.cs.cmu.edu/classes/95-865-K/HW/HW3/epinions3.zip`.

The aforementioned dataset has both positive and negative feedback about a product, as shown in the following screenshot:

Neg	Our families have always bought Ford cars and trucks We have always been treated good until now My husband and I love
Neg	I am the victim of a 2001 Ford Focus ZTS It s tragic that it is such a pleasant car to drive How could a nice car to drive be
Neg	I purchased a 1994 Ford Probe SE in April 1999 with 67 000 miles on the odometer for 5 300 It now has about 87 000 mile
Neg	I bought a 1994 Probe GT with only 400 miles on it for 22 000 I took very good care of my car servicing it often Therefore
Pos	Some of you might have read my previous post on the Ford Taurus Some might not have Right now I m at my wit s end I t
Pos	I bought my new 1999 Ford Taurus just two months ago but I m in love My former car a 91 Mercury Sable had given me
Pos	My first car was a Ford One of those old clunky ugly suckers It ran like it looked My second new car also was a Ford OK
Pos	Last spring we got a new car a 1999 ford taurus We are very happy with it Why wouldn t we be We got a great deal from
Pos	Recently I bought a used 99 Ford Taurus The model came with a V6 engine power lumbar seat cruise control a nice pack

We will train the dataset using the scikit-learn pipeline and LinearSVC. Let's take a closer look at both of these.

Scikit-learn

This is a data mining and data analysis Python library built on top of **NumPy**, **SciPy**, and **Matplotlib**. This helps with ML problems related to classification, regression, clustering, and dimensionality reduction.

The scikit-learn pipeline

The main purpose of the scikit-learn pipeline is to assemble ML steps. This can be cross-validated to set various parameters. Scikit-learn provides a library of transformers that are used for preprocessing data (data cleaning), kernel approximation (expand), unsupervised dimensionality reduction (reduce), and feature extraction (generate). The pipeline contains a series of transformers with a final estimator.

The pipeline sequentially applies a list of transforms, followed by a final estimator. In the pipeline, the `fit` and `transform` methods are implemented during the intermediate steps. The `fit` method is implemented only at the end of pipeline operation by the final estimator. To cache the transformers in the pipeline, memory arguments are used.

An estimator for classification is a Python object that implements the method's fit (*x*, *y*) and predict (*T*) values. An example of this is `class sklearn.svm.SVC`, which implements SVC. The model's parameters are taken as arguments for the estimator's constructor. The `memory` class in scikit-learn has the `class sklearn.utils.Memory(*args, **kwargs)` signature. This has methods to cache, clear, reduce, evaluate, and format the memory objects. The `cache` method is used to compute the return value of the function. The returned object is a `MemorizedFunc` object, which behaves like a function and offers additional methods for cache lookup and management. The `cache` method takes parameters such as `func=None, ignore=None, verbose=None,` and `mmap_mode=False`.

The `class signature` pipeline is as follows:

```
class sklearn.pipeline.Pipeline(steps, memory=None)
```

Let's take a look at another important component in the next section.

LinearSVC

One of the classes in the scikit-learn library is LinearSVC, which supports both sparse and dense types of input. A one-versus-the-rest scheme is used to handle the multiclass support. LinearSVC is similar to SVC, where the parameter is `kernel = linear`, but `liblinear` is used to implement the parameter in LinearSVC, rather than `libvsm`, which is used in SVC. This provides us with more flexibility to choose the penalties and loss functions. It also helps in scaling a large number of samples in a better way.

The `class` signature is as follows:

```
class sklearn.svm.LinearSVC(penalty='l2', loss='squared_hinge', dual=True,
tol=0.0001, C=1.0, multi_class='ovr', fit_intercept=True,
intercept_scaling=1, class_weight=None, verbose=0, random_state=None,
max_iter=1000)
```

It's now time to start building our model, as follows:

1. We will start by importing all the necessary libraries, as follows:

```
import re
import coremltools
import pandas as pd
import numpy as np
```

```
from nltk.corpus import stopwords
from nltk import word_tokenize
from string import punctuation
from sklearn.feature_extraction import DictVectorizer
from sklearn.pipeline import Pipeline
from sklearn.svm import LinearSVC
from sklearn.model_selection import GridSearchCV
```

The `re` library is the regular expressions library that provides the matching operations that make it easy to handle text data. The `nltk` library is used to format the text according to our requirements, while `sklearn` offers the ML tools required. The `coremltools` library helps us in to convert the `sklearn` model to a Core ML model.

2. Now, let's start reading our input, as follows:

```
# Read reviews from CSV
reviews = pd.read_csv('epinions.csv')
reviews = reviews.as_matrix()[:, :]
print "%d reviews in dataset" % len(reviews)
```

The preceding code reads the CSV file and then converts it into a `numpy` array that includes all the rows and columns. Now that we have the dataset ready, we can start extracting the features from the data.

3. Now, let's work on feature selection, as follows:

```
# Create features
def features(sentence):
 stop_words = stopwords.words('english') + list(punctuation)
 words = word_tokenize(sentence)
 words = [w.lower() for w in words]
 filtered = [w for w in words if w not in stop_words and not
             w.isdigit()]
 words = {}
 for word in filtered:
     if word in words:
         words[word] += 1.0
     else:
         words[word] = 1.0
 return words
```

4. We will start by vectorizing the `features` function. Then, we will extract the features of every sentence in the DataFrame and store them in an X variable. After this, we will set the target variable. The target variable is going to be the output. In our case, we will get a label for every sentence that indicates the sentiment in it:

```
# Vectorize the features function
features = np.vectorize(features)
# Extract the features for the whole dataset
X = features(reviews[:, 1])
# Set the targets
y = reviews[:, 0]
```

5. In our case, we will create a pipeline with `DictVectorizer` and `LinearSVC`. `DictVectorizer`, as the name suggests, converts the dictionary in to vectors. We have picked `GridSearchCV` to select the best model from a family of models, parametrized by a grid of parameters:

```
# Do grid search
clf = Pipeline([("dct", DictVectorizer()), ("svc", LinearSVC())])
params = {
  "svc__C": [1e15, 1e13, 1e11, 1e9, 1e7, 1e5, 1e3, 1e1, 1e-1, 1e-3,
            1e-5]
}
gs = GridSearchCV(clf, params, cv=10, verbose=2, n_jobs=-1)
gs.fit(X, y)
model = gs.best_estimator_
```

6. We will then print the results, as follows:

```
# Print results
print model.score(X, y)
print "Optimized parameters: ", model
print "Best CV score: ", gs.bestscore
```

7. We can now convert the scikit-learn model into `mlmodel`, as follows:

```
# Convert to CoreML model
coreml_model = coremltools.converters.sklearn.convert(model)
coreml_model.short_description = 'Sentiment analysis AI projects.'
coreml_model.input_description['input'] = 'Features extracted from
                                    the text.'
coreml_model.output_description['classLabel'] = 'The most likely
polarity (positive or negative or neutral), for the given input.'
coreml_model.output_description['classProbability'] = 'The
probabilities for each class label, for the given input.'
coreml_model.save('Sentiment.mlmodel')
```

Once we have our model, we can start building the application.

Building the iOS application

Let's start building the iOS application with the model that was built in the previous step. The model will predict the output according to whether the input text is positive, neutral, or negative in nature.

To build this application, Xcode version 10.1 should be used:

1. Create a new project with a **Single View app**, as illustrated in the following screenshot:

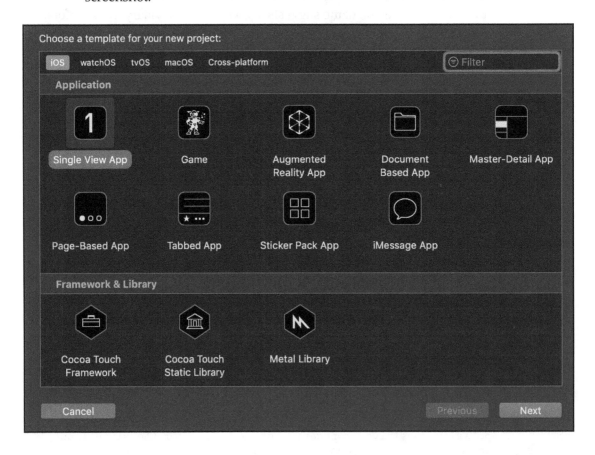

2. Mention the name of our application on the **Next** screen.
3. On the next wizard screen, pick an appropriate name for your application.
4. Fill in the rest of the fields, including **Organization Name**, as well as **Organization Identifier**.
5. We are not going to use core data in this application, so let's skip that option.
6. Let's start by creating a new app in Xcode. The following screenshot demonstrates how to create a new project in Xcode:

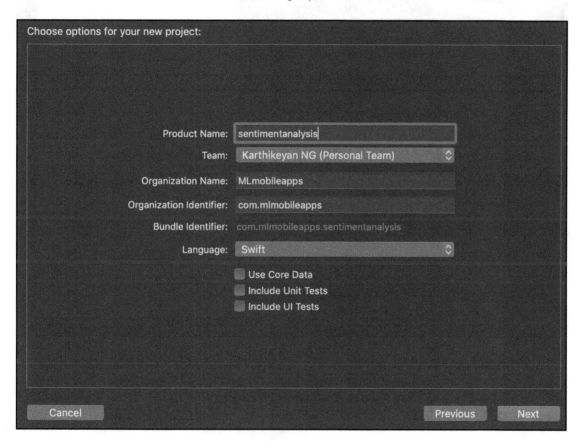

7. Next, create a storyboard, as shown in the following screenshot:

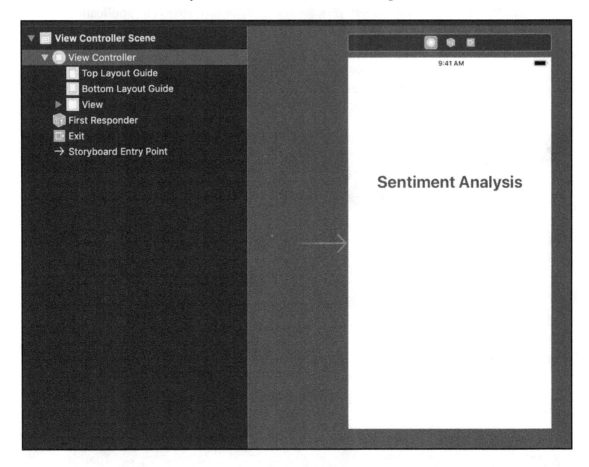

8. Once you select the file location in which to save your application, you will be able to see the **General** tab with information on the new application that has been initialized, as shown in the following screenshot:

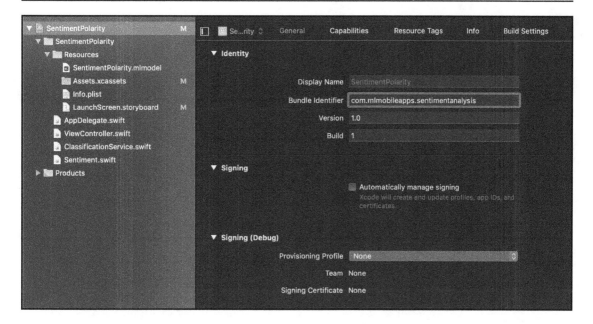

9. We will create a simple UI with a button at the bottom to display the sentiment:

```
//initialize Ui components
private lazy var textView: UITextView = self.makeTextView()
private lazy var accessoryView = UIView()
private lazy var resultLabel: UILabel = self.makeResultLabel()
private lazy var clearButton: UIButton = self.makeClearButton()
private let padding = CGFloat(16)
private var textViewBottomConstraint: NSLayoutConstraint?
```

10. We will define `sentiments` as the enumerator, as follows:

```
enum Sentiment {
    case neutral
    case positive
    case negative
    var emoji: String {
        switch self {
        case .neutral:
            return "😐"
        case .positive:
            return "😊"
        case .negative:
            return "😞"
    }
```

```
        }
      var color: UIColor? {
          switch self {
              case .neutral:
              return UIColor(named: "NeutralColor")
              case .positive:
              return UIColor(named: "PositiveColor")
              case .negative:
              return UIColor(named: "NegativeColor")
          }
      }
  }
```

11. Let's write the `ClassificationService` to fetch the result from the model that we have built:

```
//variables initialization
private let options: NSLinguisticTagger.Options =
[.omitWhitespace, .omitPunctuation, .omitOther]
private lazy var tagger: NSLinguisticTagger = .init(
  tagSchemes: NSLinguisticTagger.availableTagSchemes(forLanguage:
"en"),
  options: Int(self.options.rawValue)
)

private extension ClassificationService {
 func features(from text: String) -> [String: Double] {
    var wordCounts = [String: Double]()
    tagger.string = text
    let range = NSRange(location: 0, length: text.utf16.count)
    // let's tokenize the input text and count the sentence
    tagger.enumerateTags(in: range, scheme: .nameType, options:
options) { _, tokenRange, _, _ in
        let token = (text as NSString).substring(with:
tokenRange).lowercased()
        // Skip small words
        guard token.count >= 3 else {
        return
        }
        if let value = wordCounts[token] {
            wordCounts[token] = value + 1.0
        } else {
            wordCounts[token] = 1.0
        }
    }
 return wordCounts
 }
```

12. The input is passed on to the `prediction` method to filter the statements into positive, negative, or neutral **sentiments**:

```
func predictSentiment(from text: String) -> Sentiment {
do {
let inputFeatures = features(from: text)
// Make prediction only with 2 or more words
guard inputFeatures.count > 1 else {
throw Error.featuresMissing
}
let output = try model.prediction(input: inputFeatures)
    switch output.classLabel {
        case "Positive":
            return .positive
        case "Negative":
            return .negative
        default:
            return .neutral
            }
} catch {
    return .neutral
    }
}
}
```

13. Let's write `ViewController` by initializing the `view` components, as follows:

```
override func viewDidLoad() {
 super.viewDidLoad()
 title = "Sentiment Analysis".uppercased()
 view.backgroundColor = UIColor(named: "BackgroundColor")
 view.addSubview(textView)
 accessoryView.frame = CGRect(x: 0, y: 0, width:
view.frame.size.width, height: 60)
 accessoryView.addSubview(resultLabel)
 accessoryView.addSubview(clearButton)
 textView.inputAccessoryView = accessoryView
NotificationCenter.default.addObserver(
 self,
 selector: #selector(keyboardDidShow(notification:)),
 name: .UIKeyboardDidShow,
 object: nil
 )
 setupConstraints()
 //Show default sentiment as neutral
 show(sentiment: .neutral)
 }
```

14. The initial `setupConstraints` on the buttons and labels are defined as follows:

```
func setupConstraints() {
 //input textview
 textView.translatesAutoresizingMaskIntoConstraints = false
 textView.topAnchor.constraint(equalTo: view.topAnchor, constant:
80).isActive = true
 textView.leadingAnchor.constraint(equalTo: view.leadingAnchor,
constant: padding).isActive = true
 textView.trailingAnchor.constraint(equalTo: view.trailingAnchor,
constant: -padding).isActive = true
 textViewBottomConstraint =
textView.bottomAnchor.constraint(equalTo: view.bottomAnchor)
 textViewBottomConstraint?.isActive = true

 //result label at the bottom
 resultLabel.translatesAutoresizingMaskIntoConstraints = false
 resultLabel.centerXAnchor.constraint(equalTo:
accessoryView.centerXAnchor).isActive = true
 resultLabel.centerYAnchor.constraint(equalTo:
accessoryView.centerYAnchor).isActive = true

 //Clear button at the bottom right
 clearButton.translatesAutoresizingMaskIntoConstraints = false
 clearButton.trailingAnchor.constraint(
 equalTo: accessoryView.trailingAnchor,
 constant: -padding
 ).isActive = true
 clearButton.centerYAnchor.constraint(equalTo:
accessoryView.centerYAnchor).isActive = true
 }
```

15. Define the `Show()` method, as follows:

```
private func show(sentiment: Sentiment) {
accessoryView.backgroundColor = sentiment.color
        resultLabel.text = sentiment.emoji
 }
```

16. Let's run the application on the simulator. You can see the output in the following screenshot:

17. Now, let's use different inputs for our application and get the output, as follows:

18. An example statement of a negative input with the same output is shown in the following screenshot:

19. An example input using neutral text is shown in the following screenshot:

Here, we are able to get the sentiment from the given text input. Now, you can go one step further by improving the existing model.

You can explore finding sentiments on images in further detail in various sources. An example application is *Fine-tuning CNNs for Visual Sentiment Prediction*. You can read about this application at `https://github.com/imatge-upc/sentiment-2017-imavis`.

Summary

At this point, you should be able to build your own TensorFlow model and convert it into a Core ML model so that it can be used in an iOS application. The same TensorFlow model can be converted into a TensorFlow Lite model, which can then be used in an Android application or iOS application. Now, you can take on this task and experiment with the results. That said, we are now ready to move on to the next chapter.

In the next chapter, you will use the knowledge we have acquired in this book to move on and explore how you can build your own application.

10
What is Next?

Computers are developing more and more day by day, and the form factors of devices are changing tremendously. In the past, we would only see computers in offices; however, now we see them on our home desks, on our laps, in our pockets, and on our wrists. The market is becoming increasingly varied as machines are equipped with more and more intelligence.

Almost every adult currently carries a device with them, and it is estimated that we look at our smartphones at least 50 times a day, whether there is a need for it or not. These machines affect our daily decision-making processes. Devices are now equipped with applications such as Siri, Google Assistant, Alexa, or Cortana—these are features that are designed to mimic human intelligence. The ability to answer any query thrown at them presents these types of technology as superior to humans. On the backend of this, these systems continuously improve by using the collective intelligence that is acquired from all users. The more you interact with virtual assistants, the better the results they give out.

Despite these advancements, how much closer are we to actually creating a human brain through a machine? We are in 2019 now; if science discovers a way to control the neurons of our brain, this may be possible in the near future. Machines that mimic the capabilities of a human are helping to solve complex textual, visual, and audio problems. They resemble the tasks carried out by a human brain on a daily basis; to put this into perspective, on average, the human brain makes approximately 35,000 decisions in a day.

While we will be able to mimic the human brain in the future, it will come at a cost. We don't have a cheaper solution for it at the moment. The magnitude of power consumption of a human brain simulation limits the development efforts in comparison to an actual human brain. The human brain consumes about 20 W of power, while a simulation program consumes about 1 MW of power or more. Neurons in the human brain operate at a speed of 200 Hz, while a typical microprocessor operates at a speed of 2 GHz, which is 10 million times more than the speed of neurons in the human brain.

While we are still far from cloning a human brain, we can implement an algorithm that makes conscious decisions based on previous data, as well as data from similar devices. This is where the subset of **Artificial Intelligence** (**AI**) comes in handy. With predefined algorithms that identify patterns from the complex data we have, these types of intelligence can then give us useful information.

When the computer starts making decisions without being instructed explicitly every time, we achieve **Machine Learning** (**ML**) capability. ML is used everywhere right now, including through features such as identifying email spam, recommending the best product to buy on an e-commerce website, tagging your face automatically on a social media photograph, and more. All of these are done using patterns identified in historical data, and also through algorithms that reduce unnecessary noise from the data and produce quality output. When the data accumulates more and more, the computers can make better decisions.

Mobile phones have become the default consumption medium for most of the digital products that are being produced today. As data consumption increases, we have to get results to the user as soon as possible. For example, when you scroll through your Facebook feed, a lot of content will be based on your interests and what your friends have liked. Since the time that users spend on these apps is limited, there are a lot of algorithms running on the server and client side, to load and organize content based on what you prefer to see on your Facebook feed. If there was the possibility of running all the algorithms on the local device itself, we wouldn't need to depend on the internet to load the content faster. This is only possible by performing the processes on the client's device itself, instead of processing in the cloud.

As the processing capability of mobile devices increases, we will be encouraged to run all ML models on the mobile device itself. There are a lot of services that are already being processed on the client's device, such as identifying a face from a photo (such as Apple's Face ID feature), which uses ML on the local device.

While multiple topics are trending such as AI, **augmented reality** (**AR**), **virtual reality** (**VR**), ML, and blockchain—ML is growing faster than others, with proper use cases evident across all sectors. ML algorithms are being applied to images, text, audio, and video in order to get the output that we desire.

If you are a beginner, then there are multiple ways to start your work, by utilizing all of the free and open source frameworks that are being built. If you are worried about building a model yourself, you can start with ML Kit for Firebase from Google. Alternatively, you can build your own model using TensorFlow, and convert that model into a TensorFlow Lite model (for Android), and a Core ML model (for iOS).

In this chapter, we will cover the following topics:

- Popular ML-based cloud services
- Where to start when you build your first ML-based mobile app
- References to further reading

Popular ML–based cloud services

To begin your ML journey, you can use one of the existing cloud-based services on ML. **ML as a Service** (**MLaaS**) is widely used across all contemporary business sectors. Data is becoming cheaper, and data volume is growing exponentially. As a result, the processing power of devices is increasing at a much quicker rate. This trend has made way for multiple cloud-based services, such as **Software as a Service** (**SaaS**), **Platform as a Service** (**PaaS**), and **Infrastructure as a Service** (**IaaS**), now joined by MLaaS.

Even though we can run an ML model on our mobile device, it is still greatly impacted by limited memory, CPU and **Graphics Processing Unit** (**GPU**) resources. In this case, a cloud service comes in handy.

Starting with ML in the cloud, there are multiple services available, such as facial recognition, optical character recognition, image recognition, landmark recognition, data visualization, and **natural language processing** (**NLP**). All of these options are supported by deep neural networks, **convolutional neural networks** (**CNNs**), probabilistic models, and more. Most cloud providers run a business model that offers some free limits for a developer to explore, before deciding which is the best fit to develop their application.

The following sections will explain the four major services that are available now, and are also popular among developers and enterprises. As a beginner, you can explore the functionality under each provider and pick the one that suits your application.

IBM Watson services

IBM Watson provides deep learning as a service through a variety of products. There is a textbot service, called AI assistant, which supports mobile platforms and chat services; and a service called Watson Studio, which is helpful for building and analyzing the model. IBM Watson also has another separate API service, which processes text, vision, and speech.

There is a sample application available for developing a vision application using Core ML. This can be found at `https://github.com/watson-developer-cloud/visual-recognition-coreml`.

Microsoft Azure Cognitive Services

Microsoft provides out-of-the-box Cognitive Services in five categories, as follows:

- Vision APIs
- Speech APIs
- Knowledge APIs
- Search APIs
- Language APIs

The function of these APIs is mentioned in the following sections.

Vision APIs

Vision APIs are algorithms used for image processing to recognize, caption, and moderate your pictures smartly.

Speech APIs

Through speech APIs, the spoken audio is converted to text. The APIs use voice print for verification, or for adding speaker recognition to an application.

Knowledge APIs

To solve tasks such as intelligent recommendations and semantic searches, we use knowledge APIs that help us to map the complex information and data.

Search APIs

Search APIs provide Bing web search APIs to your apps. They also harnesses the ability to combine billions of web pages, images, videos, and news with a single API call.

Language APIs

Language APIs allow your apps to process natural language with pre-built scripts, evaluate sentiments, and learn how to recognize what the user needs.

> There are multiple sample applications for the preceding APIs. These can be found at `https://azure.microsoft.com/en-in/resources/samples/?%20sort=0sort=0`.

Amazon ML

Amazon Web Services (**AWS**) has multiple offerings for ML-based services. All of these services are tightly coupled, in order to work efficiently in the AWS cloud. A few of the services are highlighted in the following sections.

Vision services

AWS has Amazon Rekognition, which is a deep learning-based service that is designed to process images and videos. We can integrate the service on mobile devices as well.

Chat services

Amazon Lex helps to build chatbots. This industry is still growing, with more and more data coming in; the service will become more intelligent, allowing it to answer queries even better.

Language services

Some examples of language services include Amazon Comprehend, which helps to uncover insights and relationships in text; Amazon Translate, which helps with the fluent translation of text; Amazon Transcribe, which helps with automatic speech recognition; and Amazon Polly, which helps to turn natural-sounding text into speech.

You can see a few sample applications at `https://github.com/aws-samples/machine-learning-samples`.

Google Cloud ML

If you want to run your model in the cloud, the Google Cloud ML Engine offers the power and flexibility of TensorFlow, scikit-learn, and XGBoost in the cloud. If this is not suitable, you can pick the API services that best fit your scenario. In Google Cloud ML, multiple APIs are available. These are classified into four major categories, as follows:

- **Vision**: The Cloud Vision API helps with image recognition and classification; the Cloud Video Intelligence API helps with scene-level video annotation; and AutoML Vision helps with custom image classification models.
- **Conversation**: Dialogflow Enterprise Edition builds conversational interfaces; the Cloud Text-to-Speech API converts text to speech; and the Cloud Speech-to-Text API converts speech to text.
- **Language**: The Cloud Translation API is used in language detection and translation; the Cloud Natural Language API is used in text parsing and analysis; AutoML Translation is used in custom domain-specific translation; and AutoML Natural Language helps in building custom text classification models.
- **Knowledge**: The Cloud Inference API derives insights from time series datasets.

You can find a number of Google Vision APIs at `https://github.com/GoogleCloudPlatform/cloud-vision`.

There are also other services that are popular with developers, including **Dialogflow** and **Wit.ai**.

Building your first ML model

With the knowledge that you have gained from this book, you can start to develop your own model that runs on a mobile phone. You will need to identify the problem statement first. There are many use cases where you will not need an ML model; we can't unnecessarily force ML into everything. Consequently, you need to follow a step-by-step approach before you build your own model:

1. Identify the problem.
2. Plan the effectiveness of your model; decide whether the data could be useful in predicting the output for future, similar cases. For example, collecting the purchase history for people of a similar age, gender, and location will be helpful in predicting a new customer's purchasing preferences. However, the data won't be helpful in predicting the height of a new customer, if that is the data that you are looking for.
3. Develop a simple model (this can be based on SQL). This will be useful for reducing the effort when building actual models.
4. Validate the data and throw the unnecessary data out.
5. Build the actual model.

As data is growing exponentially across various parameters (data from multiple sensors), on the local devices as well as with cloud providers, we can build better use cases with more and more personalized content. There are many applications that are already using ML, such as mail services (Gmail and Outlook) and cab services (Uber, Ola, and Lyft).

The limitations of building your own model

While ML is getting popular, it is not yet feasible to run ML models on mobile platforms to reach the masses. When you are building your own model for mobile apps, there are some limitations as well. While it is possible to make predictions on a local device without a cloud service, it is not advisable to build an evolving model that makes predictions based on your current actions and accumulates data on the local device itself. As of right now, we can run pre-built models and get inferences out of them on mobile devices, due to the constraints on memory and the processing power of the mobile devices. Once we have better processors on mobile devices, we can train and improve the model on the local device.

There are a lot of use cases related to this. Apple's Face ID is one such example, running a model on a local device that requires computations from a CPU or GPU. When the device's capability increases in the future, it will be possible to build a completely new model on the device itself.

Accuracy is another reason why people refrain from developing models on their mobile devices. Since we are currently unable to run heavy operations on our mobile devices, the accuracy, as compared to a cloud-based service, seems bleak, the reason for this being the limitations on both memory and computational capability. You could run the models that are available for mobile devices in the TensorFlow and Core ML libraries instead.

The TensorFlow Lite models can be found at `https://www.tensorflow.org/lite/models`; and the Core ML models can be found at `https://github.com/likedan/Awesome-CoreML-Models`.

Personalized user experience

A personalized **user experience** (**UX**) will be the basic use case for any mobile based consumer business, to provide a more curated and personalized experience for the users of their mobile applications. This can be done by analyzing data points, such as the following:

- Who is your customer?
- Where do they come from?
- What are their interests?
- What do they say about you?
- Where did they find you?
- Do they have any pain points?
- Can they afford your products?
- Do they have a history of purchases or searches?

For example, consider the customer of a retail company or a restaurant. If we have answers to the preceding questions, we have rich data about the customers, from which we can build data models that will provide more personalized experiences (with the help of ML). We can analyze and identify similar customer, to provide a better UX for all of the users, as well as targeting the right future customers.

Let's take a look at these ideas in further detail in the following sections.

Providing better search results

Providing better search results is one of the major use cases, especially on a mobile application, to provide more contextual results, rather than text-based results. This will help to improve the business's customer base. ML algorithms should learn from user searches and optimize the results. Even spelling corrections can be done intuitively. Moreover, collecting the user's behavioral data (concerning how they use your app) will be helpful in providing the best search results, so that you can rank the results in a way that is personalized to the user.

Targeting the right user

Most apps capture the user's age and gender data when they install the application for the first time. This will help you to understand the common user group of your application. You will also have the user's data, which will give you the usage and frequency of how much the user utilizes the app, as well as location data, if that is permitted from the user's end. This will be helpful in predicting customer targets in the near future. For example, you will be able to see whether your user audience is coming from an age group of between 18 and 25, and is predominantly female. In that case, you could devise a strategy to pull more male users, or just stick to targeting female users only. The algorithm should be able to predict and analyze all of this data, which will be helpful in marketing to and increasing your user base.

There are a lot of niche use cases where ML-based mobile apps can be of great help; some of them are as follows:

- Automatic product tagging
- Time estimations, as used in Pedometer, Uber, and Lyft
- Health-based recommendations
- Shipping cost estimations
- Supply chain predictions
- Money management
- Logistics optimization
- Increasing productivity

Summary

With all the basic ideas that you have gained from this book, you can start building your own application with ML capabilities. Furthermore, with all of the new ways to interact with devices such as Amazon Alexa, Google Home, or the Facebook portal, you will find more use cases to build ML applications. Ultimately, we are moving toward a world with more connected devices, bringing the connections and communications closer to us, and leading us to create better experiences with ML.

Further reading

There are a lot of ML courses that are available online. A few of these courses are listed as follows:

- If you are beginner, you can start with the Coursera tutorial on ML by Andrew Ng, which can be found at `https://www.coursera.org/learn/machine-learning`
- An ML crash course from Google can be found at `https://developers.google.com/machine-learning/crash-course/`
- One of the best (and most enlightening) ML based blog series, by Adam Geitgey, can be found at `https://medium.comi@ageitgey/machine-learning-is-fun-80ea3ec3c471`
- You can kick-start your skills in TensorFlow at `https://codelabs.developers.google.com/codelabs/tensorflow-for-poets`
- A more thorough look at TensorFlow can be found at `https://petewarden.com/2016/09/27/tensorflow-for-mobile-poets/`

Other Books You May Enjoy

If you enjoyed this book, you may be interested in these other books by Packt:

Artificial Intelligence for Big Data
Anand Deshpande, Manish Kumar

ISBN: 9781788472173

- Manage Artificial Intelligence techniques for big data with Java
- Build smart systems to analyze data for enhanced customer experience
- Learn to use Artificial Intelligence frameworks for big data
- Understand complex problems with algorithms and Neuro-Fuzzy systems
- Design stratagems to leverage data using Machine Learning process
- Apply Deep Learning techniques to prepare data for modeling
- Construct models that learn from data using open source tools
- Analyze big data problems using scalable Machine Learning algorithms

Hands-On Artificial Intelligence for Beginners
Patrick D. Smith

ISBN: 9781788991063

- Use TensorFlow packages to create AI systems
- Build feedforward, convolutional, and recurrent neural networks
- Implement generative models for text generation
- Build reinforcement learning algorithms to play games
- Assemble RNNs, CNNs, and decoders to create an intelligent assistant
- Utilize RNNs to predict stock market behavior
- Create and scale training pipelines and deployment architectures for AI systems

Leave a review - let other readers know what you think

Please share your thoughts on this book with others by leaving a review on the site that you bought it from. If you purchased the book from Amazon, please leave us an honest review on this book's Amazon page. This is vital so that other potential readers can see and use your unbiased opinion to make purchasing decisions, we can understand what our customers think about our products, and our authors can see your feedback on the title that they have worked with Packt to create. It will only take a few minutes of your time, but is valuable to other potential customers, our authors, and Packt. Thank you!

Index

www.ingramcontent.com/pod-product-compliance
Lightning Source LLC
Chambersburg PA
CBHW080625060326
40690CB00021B/4823